Introduction to Computer Fundamentals

Bright Siaw Afriyie

Trafford rev. 08/14/2018

www.trafford.com

North America & international
toll-free: 1 888 232 4444 (USA & Canada)
fax: 812 355 4082

Preface

As the title suggests, this book is intended for those who want to learn about computers. It's a useful tool for both beginners and advanced computer users. The main reason behind the publishing of this book is to offer a little contribution to a concise learning material to boost computer literacy among the mass population that have some difficulties in coping up with the fast pace of the endless computer evolution. To some, it may sound weird, but the undeniable fact is that even if you're an expert in the field you would still have to be on your toes to be able to catch up with the fast growing pace of today's computer technology. It will therefore make sense to gain a prior exposure to some of the vital fundamental concepts covered in this book.

I want to ask you an unavoidable question, "Why would anyone want to use a computer". The answer is simple: to enhance any business activity. This new millennium is practically information technology era. Almost no efficient work could be done without the computer power. There is greater need for any user, being an individual or a business, to know which computers to use for a particular job, which implies that appropriate specifications must be considered before even deciding to purchase.

This book contains useful information that both beginners and advanced users need to know to lay a solid foundation for computing - embracing hardware and software components. This book provides an introduction to computer studies with the objective of guiding readers from hardware components through software interface. In this book, readers will learn much about the computer machine. After reading this book thoroughly, readers may not worry about the need to consult other sources to understand how computer works, but would have more confidence in computer business ranging from assembling components, maintenance to basic system designing and computer marketing. The main goal here is to acquaint you with the main hardware and software components – enough so you would be able to continue your passion in computer education on your own.

It has been my experience that introductory computer science books tend to do a great deal of handholding. They cover concepts at a very slow pace, primarily by padding them heavily with overblown examples and irrelevant junk, that only the author could conceivably find amusing. In my opinion, this seems to be unprofessional and the reader would definitely find it quite disadvantageous to take perhaps 320 pages and pad it with 600 pages of irrelevant junk.

This book is straightforward and concise to the subject matter. From chapter one of getting started through chapter eight of hardware software, the reader would have a lot to discover from the 320 pages in the present edition.

To all beginners chapter one would perhaps be the solid concrete. It discusses not only the computer evolution, but also removes some of the fear new beginners might have in operating computers for the first time.

The Book's Audience

As an Introduction to Computer Studies, the book is primarily addressed to two groups of readers;
- Beginners and,
- Advanced Users

Beginners group: The beginners group includes first time computer users, students reading computer science and information technology, computer sales representatives and new marketing agents. It's also for any one who would like to know more about the in-and-outs of the computer machine.

Advanced group: The advanced group includes advanced computer users, computer technicians, students reading computer science and engineering. System Engineers, and developers need accurate computer specs analysis in designing in order to determine the hardware and software requirements. Programmers also fall into this category. Computer experts can also take full advantage of this book by using it as a reference manual and a companion handbook.

Organization of this Book
Introduction to computer studies is composed of eight chapters with a glossary at the end. The organization of this book follows the principles of modern designing methodology commencing with simple abstraction, and gradually walking readers through more advanced stuffs without feeling the pinch.

Chapter 1. *Getting started Now*, talks about the history behind the invention of computer machine. It examines the types of computers, the evolution of the computer systems and identifies the major aspects of the primitive architectures currently being used in modern computer technology.

Chapter 2. *Computer Hardware – Motherboard*, introduces the general overview of computer hardware. It describes the general system architecture laying more emphasis on the computer main-board (motherboard). We identify all the major system components in this chapter and put them together as we read through the subsequent chapters.

Chapter 3. *Computer Hardware – CPU*, introduces the reader to the structure of the microprocessor, CPU. It examines the CPU performance, speed, types and principles of operation. The combination of lesson materials compiled in chapters two and three set the tone leading to a better understanding of the functions of lower layer programs like system drivers and assembler language.

Chapter 4. *Computer Memory*, examines the types of the computer memory. It introduces Caching and Virtual memory concepts, as well as upgrading the computer system using memory chips.

Chapter 5. *Computer Permanent Memory*, introduces the permanent storage devices. It covers all types of the computer permanent memories, their structure, seeking and rotational speeds. This chapter also discusses the mechanism of operation of computer storage devices. The coverage of storage memories and disk management in both chapters four and five identifies the key elements needed in preparing readers to digest the principles of operation of the operating system, without which no modern computer will function. Having completed these chapters the reader will now be ready to make good sense of the material covered in the Chapters 6 and 7.

Chapter 6. *Computer Peripherals*, introduces both internal and external components that surrounds the CPU system. How these devices can communicate with the system. It also examines the general input – output and identifies their corresponding unit devices such as the monitor, keyboard, mouse, printer, scanner, modem. It provides link between internal and external devices by discussing peripheral port interfaces. This chapter certainly exposes the reader to appreciate the basis of system configuration. A know-how in device conflict and resolution is a big plus for any computer user.

Chapter 7. *Computer Software and Hardware*, introduces the integration of both software and hardware. It examines the computer software and hardware interface, hardware dependency, role of bios in software hardware communications, the hardware dependency and character sets. It introduces computer program and software, types of computer software and the bridge between hardware and software making reference to BIOS, basic input-output system as the most fundamental software required for any computer to operate.

Chapter 8. *Basic Elements Computer Programming*, introduces the basic programming concepts. It examines computer systems analysis concepts including the development of algorithms, flowcharts. Readers will be introduced to the know-how in creating simple programs. It will also discuss Data Types, Variables, Constants and Arrays, Loops and Conditional decisions in programs of both software and hardware. The application of these programming concepts are covered under this chapter.

From here I hope the reader would build a solid foundation for all diversified computing work.

About the Author

Bright Siaw Afriyie is a professional Information Technology Analyst. He is also the founder of Sab Softech USA. He completed High School in Opoku Ware School in Kumasi, had a baccalaureate degree (BSc.) in Computer Science and an advanced degree in Telecommunications in University of Quebec in Canada. Bright is currently pursuing an MBA in Graduate School of Management (GSM) at University of Dallas, Texas.

Bright Siaw Afriyie worked as a programmer/statistics for the World Health Organization (WHO) Onchocerciaisis Control Program in West Africa for seven years while stationed in Ouagadougou in Burkina Faso. As a professor Assistant, Bright had taught Computer Data Communications in University of Quebec in Montreal for two years. In the United States Bright Siaw Afriyie had served as adjunct faculty in both Mountain View and Brookhaven Colleges where taught Visual Basic Programming and Object Oriented Java programming. Bright is presently working as Information Technology professional in the City of Dallas, Texas in designing and maintaining several automated systems for Dallas 9-1-1 emergency dispatch. He has created several cross-platform systems that are currently functional supporting 24/7 emergency operations. Bright Siaw Afriyie (Nana Taaka II) is also a Chief of Adansi-Atobiase, a small town in the Ashanti Region, Ghana.

Why I should buy this book?

Taking notice of the trend of industries massive migration from manual operations streamline to high-tech resolutions, as a smart clue for this new millennium readiness, this book has been prepared for you to uncover several confusing concepts that pose a big challenge to computer learners and users. I am coming from both educational and professional standpoint to better alienate the hinges that serve as obstacles to high-tech solutions to everyone. It is the togetherness of a great practical experience, educational and teaching skills, technical know-how and continuous customer value-added service research that has always been the source of creation of this book and two other computer science books yet to be published. The feedbacks so far received from few professors in Information technology in Dallas, Texas area strongly suggests the use of this book as a great fundamental and companion material for computer science students.

The organization of the core material in this book both provides support training unconditionally to every one who wants to be computer literate, and also extends its learning curve to high quality systems engineering to individuals or companies already operational in the high-tech industry.

This book provides a solid foundation for information technology. You don't want to miss this good news.

Bibliography

Computer Organization and Design. The hardware/software Interface. *John L. Hennessy and David A. Patterson. Morgan Kaufmann Publishers, Inc, Sans Francisco, California*
ISBN 1-55860-281-X

Data Communications, Computer Networks and Open Systems Fourth Edition. *Fred Halsal. Addison Wesley Publishing Company. ISBN 0-201-42293-X*

Modern Electronic Communication, Fifth Edition. *Gary M. Miller - Prentice-Hall International Publishers.*

How to do everything with your PC. *Robert Cowart – Osborne/McGraw Hill ISBN 0072127767*

The do-it-yourself PC book: An illustrated guide to upgrading and repairing your computer. *MacRae Kyle - Berkeley California Osborne/McGraw Hil l-ISBN 0072133775*

Teach yourself PC's in 24 hours. *Greg Perry M – Sams Pub. Indianapolis Ind. ISBN 0672311631.*

The first week wih my new PC: a very basic guide for mature adults and everyone else who wants to get connected. *Pamela R. Lessing ISBN 1892123223*

DEDICATION

This book is dedicated to:

1. My Son, Samuel Kofi Siaw Afriyie Jr.
2. My Dearest Wife, Lucy Siaw Afriyie for her love and support.
3. The rest of my family, and especially to my Mother and in memory of Father the Late, Samuel Kwasi Siaw

ACKNOWLEDGEMENT

I am most thankful to My Heavenly Father for His amazing grace and guidance in my designing, teaching and research work. I thank my family for their unchanging support for consistent long hours of research work and editing.

My sincere thanks go to Lt. John Settles, and William Lingburghe of Dallas Fire Rescue and all Bright Idea Solution partners in Dallas Texas, USA for their financial support in revising this second edition. I give a special thanks go to Mr. Jerry R. Martin Jr. of Emergency Preparedness of City of Dallas, Texas.

I also extend my sincere thanks to all the Dallas Fire Rescue officers who also supported me in anyway during this re-editing.

BRIGHT SIAW AFRIYIE

Table Of Contents

Chapter 3 – Computer CPU   47 -103

Chapter 4 – Computer Memory   103 -129

Chapter 5 – Storage Devices - *Permanent memory*   130 -174

Chapter 6 - Peripheral Devices   175-220

Chapter 7 - Software - Hardware Interface   222-251

Chapter 8 - Basic Elements of Programming 252-290

SAB-SOFTECH, USA
SAB TECHNOLOGIES
GHANA

Getting Started Now!

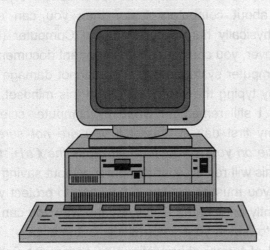

Chapter 1

Get Started Now!

What is a Computer?

Computer Types

Know your Computer Power

Evolution of the Computer

Get Started Now !

Most beginners panic when they touch a computer for the first time. Perhaps, they think typing a wrong key on the keyboard may cause some damage to the computer. Do not panic! You'll learn everything that's really important about computers faster than you can ever imagine. You cannot physically harm your Personal Computer (**PC**) from the keyboard. However, you could erase an important document or file or reorganize the computer system setup. You cannot damage the computer components by typing the wrong key. With this mindset, we are now ready to roll. I still remember what my computer science professor told me on my first day in class: *"If you are not sure of something you have done on your computer, just press the Ctrl+ Alt + Delete simultaneously.* This will restart your computer without saving the changes you made. So you must take extra precaution to protect your work by saving it frequently prior to the D-day. By the way you can set the auto-save feature to save your information automatically. The computer industry is full of terms; abbreviations or **acronyms** (letters that stand for words and sound like the words themselves). To a beginner, the biggest computer challenge is understanding what all the acronyms mean. Most often you will see acronyms like **CPU, DVD, MODEM, NIC, RAM, SCSI,** and many others in computer marketing ads and stores. This should not be a big deal. After all, if you forget any term you can refer to the glossary in the back of this book.

Computers are not only for experts. Anyone can be expert with a computer. Actually, it is a major priority for experts to make computer usage very simple for every one. After you master the terms, you'll notice that computer is a simple machine that you can use for beneficial work and fun. Anyone can be expert with computer.

What's a Computer?

The question now is how would you define a computer system to your child who is very curious about the machine. I turned round and asked my little boy, what is a computer? He answered, "*it is a smart machine that I can use to play my "Spiderman game."* A computer is simply a combination of hardware and software. The hardware represents the hard part that is actually visible to you. The part you can see. The software part is practically invisible or intangible but houses the intelligent part of the computer. You will learn a little further about hardware and software since they in effect constitute our famous computer machine. Each person will of course describe a computer system in a different way. Most people, like my little boy, have the tendency to say, "*Computers are smart*". Actually computers are not smart; people are smart. Computers are electronic circuits that sit on desks until someone tells them what to do. Even after turning on your computer, it will still require your input for it to operate. It cannot do anything and will not do anything until you tell it to do something. Your input will determine which action to take. In this way a computer interacts with people's input, and may thus be seen as a machine designed to work on the basis of human logic. It is an electronic machine capable of transmitting, storing and processing *information* or *data*.

Modern computers now have the capabilities to resolve more complex problems in a twinkle of an eye; sometimes, as fast as within a *nano-second* (*1.00 x 10^{-9}* = 1/1,000,000,000 of a *second*). The simplest description of a computer will probably be any *machine* that can manipulate or process *data*. Data comprises any set of information that can be manipulated in a computer system. The characters you type from your keyboard, mouse clicks, images and sounds you enjoy playing on your computer are all considered data. Data represents the key element, which your computer can disseminate. Data has measurable units known as *bit* and *bytes*. One byte is a collection of 8 bits. Also, **1 byte** represents the basic data unit, which the computer can interpret as a meaningful character that is readable. A bit is a binary digit equal to the number **1** or **0**. Data is fully discussed in the subsequent chapters.

Hardware and Software – Computer System

Technically, a computer system is an integrated collection of hardware and software. To be able to process information, a typical computer is designed to comprise these two specialized parts, *hardware* and *software,* which interact with each other. The hardware consists of the physical solid electronic components that are visible to you. The hardware is a *tangible* object which occupies space and has mass. Unlike the hardware, the software is a non-visible part, which is grafted to the hardware. The software part coordinates all operations that may occur in the computer. The software, the part that makes the computer look smart, constitutes, in fact, all the instructions which manipulates the hardware. Software is said to be *intangible*. It is like words printed in this book. Both are intellectual properties, and not physical entities. Software consists of programs that tell the hardware what to do.

Types and Uses Of Computers

Some computer systems are intended for direct use by humans so, they are called *Personal Computers (PC)*. Other computer systems are embedded within other objects. For example, most modern automobiles contain embedded computers that monitor fuel consumption, control braking, provide direction guides on maps, and perform other valuable tasks. These types of computers are called *embedded systems*. Since this book is focused on introduction to computers, we will be discussing hardware and software aspects that directly affect personal computers. Different computers may exist for different tasks. However, they all operate on the basis of the collective functioning of both the hardware and software. Computers may range from smaller systems or *Personal Computers (PC),* which are sometimes called *microcomputers*, to larger systems such as *Minicomputers*, *Mainframes*, and *Super-Computers*. The most expensive supercomputer in the world acts much like your PC. Larger organizations such as companies, universities, and governments usually buy larger computers because larger computers often are faster and store more information than the PCs. The difference in computers may commonly be determined in terms of processing speed, storage capacity and quality output but not in terms of intelligence. The PC world has evolved so much that every few months computer companies make their computers faster, smaller, and less expensive. Today's PCs are faster and hold more information than yesterday's mainframes.

This book is limited to only microcomputers. Nowadays, most microcomputers are built from the original framework of IBM or Apple. IBM models are popularly known as *Personal Computers* and the Apple model on the other hand is referred to as *Macintosh*. Although both kinds of PCs do similar work, their hardware differs and so does the way you use them. The large majority of PCs in use today are the PC-compatible kind. This book primarily discusses the PC-compatible computers due to their overwhelming popularity in the computer market.

Knowing your Computer Powers

Even computer novices know something about what modern computers can do. Often, however, the mystery that surrounds PCs escapes a large majority of people who have never touched a PC. At the present time, every household or office is likely to have at least a personal computer (PC) set up for business purposes or for children games. Computerization has evolved to such an extent that almost every human activity is closer to being computer related. For example in the United States, whenever banking computers go down, business activities are almost halted since bank transactions will cease to operate. Let's try to figure out why today computers have become companions in day-to-day life? The simple reason is that computers can perform not only complex *mathematical* operations, but also have *storage* memory capabilities and can respond *intelligently* as programmed by people. With the storage capability, computers make typing and printing more fun and convenient. It becomes more economical when less time is utilized for higher productivity and efficiency. This is what makes computer involvement in many business activities more beneficial. One way of justifying this benefit is to think of balancing your checkbook manually; it's often fastidious, inaccurate and more time consuming. You can imagine how tedious it would be when generating financial reports or procurement reports from a million records, and making some bank transactions such as deposits, withdrawals and others, without computer systems. Computers also play a very major role in modern medical technology in that medical equipment such as surgical and x-ray machines, laboratory, analytical instruments and others are all computer aided. Weather forecasting centers are mainly dependent on computers. You can see the role of computers in massive industrial manufacturing in both car and aircraft industries. Multimedia computers put more fun and high quality sophisticated production in the film, television and musical industries. There are several other benefits of using computers, which are not mentioned in this book, which are left as an assignment to the reader. In summary, your PC turns raw facts and figures called *data input* into meaningful information or *data output*, bringing about high efficiency. Computers follow the **I-P-O** process, i.e., follow the Input-Process-Output method to process data. The figure below illustrates the I-P-O process.

Figure 1.1. *PC turns raw facts and figures into meaningful information.*

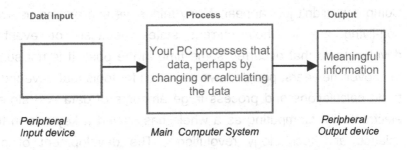

Data Input	Process	Output
	Your PC processes that data, perhaps by changing or calculating the data	Meaningful information
Peripheral Input device	*Main Computer System*	*Peripheral Output device*

The data input may come from any external device, which is attached to your PC. For example the data input may come from your telephone line, if you have phone access to another computer elsewhere, maybe via the Internet. In this case you must have a modem connected either internally or externally to your PC box. Modem stands for ***modulation-demodulation***. This device will be discussed under hardware. Your input might come from typing at your keyboard, or from a mouse click. Data input might also come from a digital camera or another computer. The PC takes the input and processes it or does something to it using the program you have installed in the computer. The output or processed information might appear on your screen, on the printer as a report or as checks, or it might even be sound and video, Modern PCs work with variety of data sources and produce all kinds of output, including *multimedia output* composed of sound, colour, and video.

The Evolution of The Computer (CPU)

Let's take a few moments to reflect on the amazing history of the computer. The headmaster of the high school I attended used to tell us " history is the wise man's subject." By that he meant we study history so we will know where we came from and how we became who we are. Computers didn't just appear. Many things we see around us living, or non-living are in their current states because of events and developments that occurred some time in the past. It is not surprising that, over the years, people have searched for tools that have the ability to do calculations and process large amounts of data and store them electronically. Computing as a whole has played a key role in today's science and technology revolution. The development of modern computers has provided the world with a new era of scientific break through, and success stories.

The history of computing and computers is an extensive field. The topics below show a few of the highlights leading to the modern concept of computing and computers. This list is not exhaustive, since new generations of computers are still being born year after year. The tremendous success centers on the significant reduction in weight and size with overwhelming increases in processing power. Computers started as machines that weighed about 30-35 tons, but today they are extremely light, weighing about 1-5 kilograms and coupled with super-powerful processing speeds.

Figure 1.2. The Computer ENIAC Compared to Modem Personal Computer.

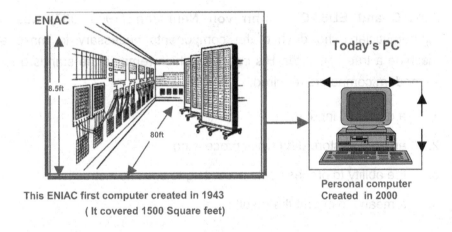

ENIAC

8.5ft

80ft

Today's PC

This ENIAC first computer created in 1943
(It covered 1500 Square feet)

Personal computer
Created in 2000

ENIAC

The first computer known as **ENIAC**, was built by J. Presper Eckert and John Mauchly at Moore School of the University of Pennsylvania during World War II. But it was publicly disclosed in **1946**. ENIAC was equipped with **18000** vacuum tubes weighed **30** tons, and measured **80** feet long and **8.5** feet high and several feet wide. It had twenty 10-digit registers, each of which was **2** feet long. ENIAC stands for Electronic Numerical Integrator And Calculator. It was created to calculate the trajectory of artillery shells. But ENIAC had two major problems. The first was the 18000 vacuum tubes, which generated a huge amount of heat that required a special cooling system. The second problem was that the program instructions had to be "hard-wired", without the flexibility to switch from one program to another. So, in order to change, program operators had to re-plug wires and reset switches. It often took users several days to program ENIAC for a problem it would solve in seconds.

EDVAC and EDSAC - John von Neumann (1903-1957) was a mathematician who defined the components necessary to make a machine a true computer. His concept, which remains the standard for all modern computers required:

1. a device to input data

2. an area to store data while processing

3. the ability to process data according to specific instructions

4. a means to output the results

Von Neumann also devised a way to encode instructions and data in the same language. Neumann actually introduced computer programming. In 1945, he wrote "***First Draft of a Report on the EDVAC***" in which he took the next step by describing *a computer which stored its program instructions in its memory* rather than punched onto tape or cards. His ideas were used in the development of

EDVAC and *EDSAC*. The stored-program model became the standard which we still use today. In 1949, Maurice Wilkes from Cambridge University improved on ENIAC and built a stored-program computer called the EDSAC (for Electronic Delay Storage Automatic Calculator). This was the world's first full-scale, operational, stored-program computer. On May 6, 1949 EDSAC ran its first program which was written by Wilkes to print a table of the squares of integers. It took EDSAC thirty seconds to print the numbers 1, 4, 9, 16, . . .n.

UNIVAC - From EDSAC, a general-purpose computer known as UNIVAC I (*Universal Automatic Computer*) was built in June 1951. UNIVAC was the first general-purpose electronic digital computer designed for commercial use. Though UNIVAC allowed data input and program using magnetic tape, operators had to use punched cards to write the programs and the initial data. Also, to make it run faster would require a special machine to translate the holes in the punched cards into magnetic data on the tape. For a program that took 120 seconds to read from a punched card, it would take only 3 seconds to read from a tape. UNIVAC was very large (about 35 tons) and very expensive. It *cost $1 million* and it was much more reliable than most earlier devices.

International Business Management (IBM) Computers
IBM/701 and System/360 - In 1952 IBM came out with the IBM/701 followed by the IBM System/360 in 1964 using the idea of the architecture abstraction in which many hardware and lower level software details were hidden from users.

Minicomputers

The first commercial *Minicomputer* model PDP-8 evolved in 1965, through the effort of the Digital Equipment Corporation (DEC). Minicomputers were the forerunners of microprocessors leading Intel to invent the microprocessors, such as Intel 4004, in 1977. In 1963 Seymour Cray announced the first Supercomputer, CDC 6000 in Minnesota. CDC stands for Control Data Corporation. And the Cray Research Inc issued Cray-1, the fastest and most expensive machine in the world.

Table 1. The illustration of the five generations of the Computer Evolution

GENERATION	DATES	TECHNOLOGY	YEAR	NAME	SIZE (CU. FT.)	MEMORY (KB)	PRINCIPAL NEW PRODUCT
1	1950-1959	Vacuum tubes	1951	UNIVAC I	1000	48	Commercial, electronic computer
2	1960-1968	Transistors	1964	IBM S360/ model 50	60	64	Cheaper computers
			1965	PDP-8	8	4	Minicomputer
3	1969-1977	Integrated circuit	1976	Cray-1	58	32,768	Minicomputers
4	1978-1990	LSI and VLSI	1981	IBM PC	1	256	Personal Computers
5	1991-20??	Microprocessor	1991	HP 9000/ Model 750	2	16,384	Personal portable computing parallel proc.

Personal Computers (PCs)

Here we are, after several decades of searching for tools that will compute and store data. The coming of this breed has allowed us the opportunity to edit this book you're now reading.

In 1977, through the effort of Steve Jobs and Steve Wozniak, the personal computer known as *Apple II*, evolved. In 1981, IBM came out with the IBM Personal Computer with **Intel 80x86**. The invention of IBM overrode the popularity of Apple's personal computers and made Intel 80x86 chip and the Microsoft Disk Operating System (**MS DOS**) the most popular chip and operating systems on the market. Operating Systems will be explained later.

A modern CPU may be equipped with a microprocessor that ranges from 8088 80-286 or 286, 80-386 or 80-486 or 486 to 80-586 or Pentium or Intel chips. At present several companies that manufacture computers include Acer, Apple, Compaq, CompUSA, Dell, Gateway, Hewllet-Packard, IBM, Micron Electronics, NEC, Packard Bell, and Sony. As of now the latest CPU microprocessor on the market at the time of this writing is Pentium IV with 1-3 GHz plus execution speed that is a higher level performance than the Pentium I, II and III.

Modern Computer Developments
Basically modern computers are made up of transistors, magnetic tapes, and integrated circuits. But the basic architecture reflects the standard defined by John von Neumann. Both hardware and software developers use the four units as the fundamental concept when building a computer system. Neumann defined the four units, which can be mapped to today's computer devices below:

1. **A device to input data:-** may now correspond to a keyboard and mouse and other input devices.
2. **An area to store data while processing:-** This is memory, cache, hard drives and registers.

3. **The ability to process data according to specific instructions:-** This refers to the modern processor or the CPU.

4. **A means to output the results:-** Devices like monitors, printers and others, that can display data into meaningful information.

This book will therefore focus on the components like motherboards, CPUs, video card, hard drive, memory and input/output devices under hardware.

Exercise 1.

Students are required to select from the five alternatives labeled A-E, the most appropriate answer to questions 1- 18.

1. Choose the sentence(s) that most define a computer.
   A. Computer is a smart machine.
   B. Electronic equipment that can transmit, store and process information.
   C. Is a system that can process information.
   D. Is a system that is composed of hardware and software.
   E. All of the above.

2. What is the difference between a computer and an ordinary calculator?
   A. Computers have efficient storage capabilities.
   B. Computers cannot calculate numbers as calculators.
   C. Calculators are usually faster and store information.
   D. Calculators are more efficient than computers.
   E. Computers are smarter than calculators.

3. What is computer hardware ?
    A. It is the computer itself.
    B. It is the physical solid electronic components.
    C. It is the intelligent part of the computer.
    D. All of B and C.
    E. None of these.

4. What is computer software?
    A. The visible part that controls the computer.
    B. The non-visible part which is grafted to the hardware.
    C. The set of instructions that coordinates all operations.
    D. All of B and C.
    E. None of these.

5. How many major types of computers exist now?
    A. Four.
    B. Two.
    C. Three.
    D. Five.
    E. Six.

6. Which of the following was the world's first computer ?
    A. UNIVAC I
    B. IBM System/360
    C. IBM 8088
    D. EDSAC
    E. ENIAC

7.   The first computer that was able to store data was called?

A. ENIAC

B. EDSAC

C. UNIVAC I

D. IBM/701

E. IBM System/360

8.   Which of the following was the first personal computer?

A. IBM System/360

B. DEC Model PDP-8

C. Apple II

D. IBM PC 8086

E. IBM PC 8080

9.   The first version of Disk Operation System (DOS) created by Microsoft came to run on Intel chip:

A. 8088

B. 8086

C. 80286

D. 80386

E. 80486

10.   Why will you choose a computer over a typewriter?

A. Because a computer types faster than typewriter.

B. Because a computer is simple to operate.

C. Because a computer is less expensive than typewriters.

D. Computers have storage capabilities and are more efficient.

E. Computers are usually smaller in size and flexible.

11. Which of the following is NOT a factor to distinguish between smaller and bigger computers?

   A. Speed
   B. Storage capacity
   C. Price
   D. Intelligence
   E. None of the above

12. To connect with the outside world via the Internet your Computer needs?

   A. Modem
   B. Cellular phone
   C. Multimedia PC
   D. Network cable
   E. None of the above

13. Which of the follow is NOT an Input device to a computer?

   A. Printer
   B. Keyboard
   C. Mouse
   D. Digital Camcorder
   E. None of the above

14. The processed information that is displayed on the monitor or is heard on a speaker is known as:

   A. Input
   B. Output
   C. Data processing
   D. Multimedia
   E. None of the above

15.   The multimedia output is composed of:

     A. Sound   D. All of these

     B. Color   E. None of these

     C. Video

16.   How many registers did the first computer have?

     A. 5   D. 20

     B. 10   E. 25

     C. 15

17.   Neumann defined a stored computer. Which of the four Components best describe this concept as applied to the major computer parts today?

     A.   Motherboard, video cards, CPU and mouse

     B.   CPU, Input devices, Output,   and Memory

     C.   CPU, Motherboard, sound cards, and keyboard

     D.   CPU, Motherboard, Monitor, and keyboard

     E.   CPU, Printer, Monitor, and keyboard

18.   The worlds first written program generated:

     A.   1, 2, 3, 4,.., n.

     B.   1, 2, 4, 8,...,n.

     C.   2, 4, 6, 8,...,n.

     D.   1, 4,9,16,...,n.

     E.   1, 3, 5, 7,...,n.

Computer Hardware - Motherboard

Chapter 2

- o **Overview of Computer Hardware**
- o **Knowing your Personal Computer Basic Parts**
- o **The Computer – System Unit**
- o **Computer Basic Architecture**
- o **Computer Basic Types Of Hardware**
- o **Main-Board Of Computer**
- o **Expansion Slots – Input/Output connectors**

Hardware Overview

Many of us use computers on a daily basis. Although we use them for many different purposes in many different ways, we share one common reason of using them; to make our job more efficient and easier. The use of personal computers has grown so much in our everyday life that it's now become necessary to acquire some basic knowledge on the computer components and how it operates.

So surprisingly, many of us don't buy this fact. They usually say, "leave that part for the computer genius to handle". How about, if you want to buy a computer for your home or business? Keep in mind, that regardless of your profession, if your job involves the use of PC's, then it's beneficial to spend a little time to read this book. It's always a good idea to know something about computers.

The word hardware refers to the physical parts of the PC that are visible to you. We now know that a computer is more than hardware. But since the hardware is the visible part it is the best starting point to learn about how computer works. A computer is a *system*, and a system is a collection of interrelated parts that work together to do something. Your PC system is not just one box but a collection of several components that operates together to analyze financial information, play games, connect to the internet, manage a business, and so on.

Knowing Personal Computer Basic Parts

Basically a personal computer hardware mainly consists of a System Unit and several accessories called *peripherals*. These peripherals in turn can be grouped into the two major categories commonly known as *input* and *output* devices. The input device refers to such peripherals as the keyboard, mouse, scanner or microphone, which may be used for entering data into the system unit. Whereas the peripherals such as monitors, printers and speakers take data out from the system unit and are therefore considered as the output devices. So for a complete circulation of data in and out, the system unit must have, in addition to a place to store inputs and outputs, the signal line that can detect when and where these operations must take place - the job of the system unit's interrupt request.

The figure 2.1 below shows a typical personal computer. Each day that passes different companies create different styles of computers, but almost every PC you see today has the standard hardware shown in figure 2.1. The speed and storage capacity of the internal components determine the price differences. The faster and more storage a PC has, the higher the price will be.

Figure 2.1. *PC consisting of a System Unit, Monitor, Keyboard, Mouse, Printer. and Speakers. A basic computer system that most people are familiar with.*

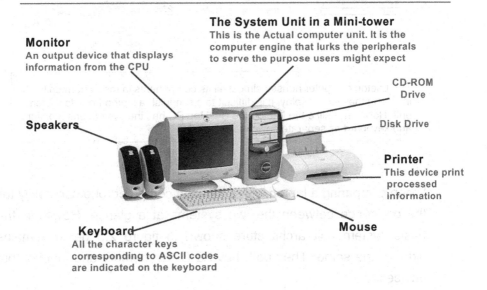

The System Unit in a Mini-tower
This is the Actual computer unit. It is the computer engine that lurks the peripherals to serve the purpose users might expect

Monitor
An output device that displays information from the CPU

CD-ROM Drive

Speakers

Disk Drive

Printer
This device print processed information

Keyboard
All the character keys corresponding to ASCII codes are indicated on the keyboard

Mouse

Many components reside inside the PC's system unit, the box that holds the disk drives, and into which the keyboard, mouse, monitor and speakers are plugged. The system unit box is also popularly known as a mini-tower when it stands upright as shown in figure 2.1. The majority of the PC's memory and storage devices reside inside the system unit. Is a laptop computer any different? Well, check it out from the two figures 2.2 and 2.3 illustrating a laptop and a desktop computers.

Figure 2.2.
Is a laptop computer

Figure 2.3
is a desktop computer

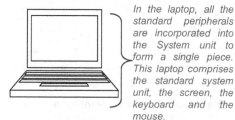

In the laptop, all the standard peripherals are incorporated into the System unit to form a single piece. This laptop comprises the standard system unit, the screen, the keyboard and the mouse.

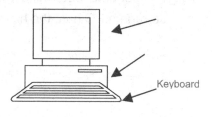

Keyboard

A Desktop Computer

The Laptop computer houses almost all its components in the system unit, including the video display. It is difficult to distinguish a laptop from its system unit. However, unlike the Laptop the Desktop PC has the system unit monitor and keyboard as separate devices.

When comparing a laptop with a desktop computer, one can easily tell the difference between the two systems at a glance. However, the basic system unit architecture shown in figure 2.5 below remains virtually the same. They both have a CPU, memory and input/output devices.

The System Unit

The system unit is the box where the brains, heart and nerve of the PC are kept. The system unit houses the computers memory, circuit boards, expansion slots, power supply and the most important component, the **Central Processing Unit** or CPU block. The CPU, which is sometimes called the **micro-processor**, is in fact, the actual computer and everything else helps the CPU. For example the act of getting data and sending data to and fro other places are considered help functions that are carried out by the peripherals. While the CPU represents the main computer block, all the remaining input/output devices surrounding it may be called **peripherals**. We can now redefine computer system as follows:

Computer = A System Unit + Peripheral Devices

Peripheral devices will be fully discussed in chapter five. However this part will set the tone to what peripheral really is. They are auxiliary devices attached to your computer allowing it to interact with its environment. Modern PC peripherals can be divided into two distinct groups: *internal* or *external*. The *internal peripherals* are components other than the CPU that is part of the system unit itself residing inside the CPU box. The *external peripherals* refer to all the other devices that may be attached externally to the CPU mini-tower or desktop as shown in figure 2.1 above. Each peripheral can also fall within one of the two functional groups, input or output. But there are exceptions to this rule. For example peripherals like modems, and network cards can do both input and output operations.

The system unit therefore simulates all these operations in the computer. All activities relating peripheral devices are processed in the system unit as input/output. A clear picture of the System unit architecture will look like what is elaborated in figure 2.5 below. The system unit can therefore be simplified by this equation:

A System unit = Processor + Memory + Input/Output

Externally the *input/outputs* are represented by the peripheral devices mentioned earlier. The Processor or CPU processes all the data that flow through your computer. The CPU may get data from an input device, such as the keyboard, to enter information into the system and an output device such as the monitor or a printer to get the results of information that had been processed in the CPU into a meaningful form that people can understand. When you open your system unit box it is most likely you might see the picture below.

Figure 2.4   A typical mini-tower housing the System Unit including the mainboard or motherboard

The System Unit (CPU)   Motherboard

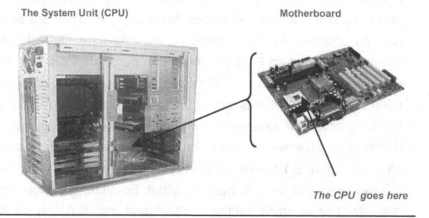

The CPU goes here

It is important to always remember that the picture above represents a computer systems unit, though in the usual technical language abuse the entire block may also be called the CPU. As had been defined earlier, the system unit may comprise not only the CPU, but also the motherboard, memory, and all the input/output devices, also described as expansion cards. Technically, a typical personal computer system may reflect one way or the other the basic computer architecture shown below in figure 2.5 that summarizes the figure 2.4 above.

Basic Architecture of Computer

In chapter one we learned about "John von Neumann's" report on how to build full-scale computer. The report was use to organize the computer system into four main parts: Central Arithmetic unit, the Central Control unit, the Memory, and the Input/Output (IO) devices.

Today, more than half a century later, nearly all processors have a "von Neumann" architecture. The internal organization of the computer system unit shown above can be grouped into the four classic components. The control and the arithmetic unit also known as the *datapath* constitute the microprocessor (CPU), and the remaining two components also collectively known as the *internal peripherals* are fixed on the motherboard to form a complete unit. This organization is illustrated in figure 2.5 below.

Figure 2.5
An Architecture of a computer

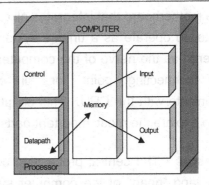

The organisation of a computer, showing the five classic components

---

Many times, a PC is said to contain several rows or banks of *integrated circuit chips*. The PC's system unit components are miniaturized to form an *integrated circuit chip*, also known as *IC*. Therefore the PC's memory and the CPU are said to be integrated circuit chips. When you add memory to your computer, you add additional *memory ICs* to the personal computer.

Basic Types of Hardware

Technically, most of the hardware in a computer system as shown in figure 2.5 may reflect the four categories: motherboard, processor, memory and peripherals (expansion cards and external peripherals devices).

The motherboard -The motherboard is the main electronic board that contains the logic circuits which tie all parts of your computer together. It provides the connectors for the CPU, the RAM, BIOS, CMOS, video cards, sound cards, storage devices, etc. Don't you bother yourself with the meaning of these terms now. They will be discussed shortly. It is the motherboard that integrates all of the components allowing the computer to operate as a unique system. The mother-board may be considered as the nerve of the computer system, by operating as the central connecting point for all computer components that communicate with the CPU. This chapter will focus on this category so as to explore the most important parts of the motherboard.

Processors - The central processing unit, or CPU, functions as the "brains" and "heart" of the computer system. The CPU is usually a single chip. A computer system may also contain additional, more-specialized processors, such as a graphics processor or a processor that performs certain types of arithmetic. The processors are called co-processors. The biggest CPU manufacturers are Intel and Motorola. Intel brands are used in IBM compatible computers, while Motorola brands are used in Macintosh computers.

Memory - Memory gives the computer the ability to remember or, store information. Basically we can identify two kinds of memory. Some types of memory store information for a relatively brief period of time, while others can store information permanently. A computer's main memory called **random-access-memory** (RAM), stores information only as long as the power is turned on. Other types of memory can store information indefinitely, even when the computer's power is turned off. The read-only memory(ROM) falls into this category. The hard disks, floppy disks, and other storage media can also store information permanently.

Peripheral devices - Peripheral devices lie on the edges of the computer system. They represent what we term as **IO** or **input/output** devices. Their job is to provide a connection, or interface (*gateway*), to the world outside the computer system. Peripheral devices include keyboard, mouse, monitor, printer, and modem.

Computer Motherboard

The motherboard is the board that contains the logic circuits that tie all parts of your PC together. You have learned that represents the nerve of the computer system unit and also provides the connectors for the CPU, the RAM, and all the internal devices. In fact any component that operates on a computer system must have a direct or indirect attachment to the motherboard in order to communicate with the CPU. The figure below is an example to show you what a motherboard looks like. Some motherboard may look different but they all share common features.

Figure 2.6

A Typical PC Motherboard showing the vital CPU, the slots for the internal peripherals and the communication ports for the external peripherals

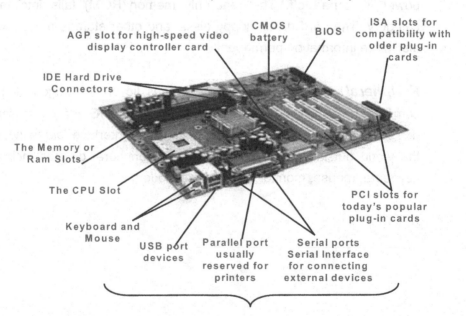

The figure 2.6 above is a picture of a motherboard that further illustrates the modern integrated circuit specifically showing the communication ports and slots for the peripherals.

The hardware architecture as illustrated in **figure 2.5** above represents the content of the mini-tower in figure 2.4 that may comprise three basic parts. The processor, memory and the **input/output** devices (I/O). Any one of the eight features can be put under one of these three components that are connected to a central board or the **motherboard**. It can also be seen as a circuit board inside the computer that provides the foundation for the computer system. The motherboard holds all the internal circuitry for the system comprising all the three parts of the system mentioned above. In every computer the motherboard is a thin green plastic, covered with dozens of small grey or black rectangles. These rectangles contain **integrated circuits** or **chips** that drives our advancing technology.

At a glance the figure above might look cumbersome with several transistors and chipsets. However to make it easier to comprehend, the complexity can be broken down into five groups. The five groups include the processors (CPU's), the memory and chipset, system BIOS(**Basic-Input-Output-System**), I/O interface, and the expansion slots. You will learn more about them individually as I walk you through this chapter. The motherboard has a whole bunch of transistors and a complex electronic circuit planted on it. A **transistor** is simply an on/off switch controlled by electricity. Thus an **integrated circuit** is a combination of dozens to hundreds of transistors to a single chip.

Motherboard Common Features

Not all motherboards are the same, but most share some common characteristics. Some of these are:

1. ISA Expansion Slots (2 long black slots at upper right)

2. PCI Expansion Slots (6 mid length white slots in middle-right)

3. AGP Expansion port (1 short brown slot in the middle)

4. CMOS battery (round disc above the PCI slots)

5. Memory (DIMM) bank (3 gray/black slots to right of SIMM slots).

   (Some have Memory (SIMM) bank (4 white slots to right of PCI slots).

6. Pentium III CPU Slot (long slot near right edge)

7. ROM Chip (BIOS Chip)

8. Clock (a chip in there somewhere)

9. 10.Input/Output ports (bottom right edge)

Motherboards come in different sizes - that is worth knowing if you are replacing an existing motherboard.

Basically most common features found on the motherboard are slots, which can accommodate internal devices. So far we have learned that peripheral devices lie on the edges of the computer system. Their job is to provide a connection, or interface (*gateway*), to the world outside the computer system. Peripheral devices include keyboards, mice, monitors, printers, scanners and modems. These peripheral devices are described in this book as external peripherals. Inside the system unit, on the motherboard there are expansion slots into which necessary expansion cards such internal modems, video, network and sound cards may slot. We may call these devices internal peripherals.

Expansion slots as available openings on the motherboard to place additional cards. They are often called *buses,* which are present in five

Figure 5.3 Mother board showing expansion slots and cards

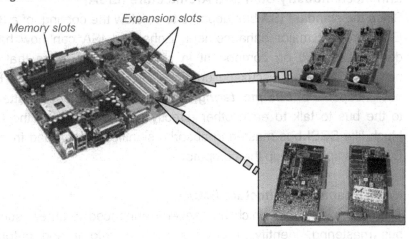

major types, *Industry Standard Architecture* (**ISA**), *Enhanced Industry Standard Architecture* (**EISA**). *Micro Channel Architecture* (**MCA**), *Video Electronics Standard Association* (**VESA**), *Personal Computer Memory Card International Association* (**PCMCIA**), *Peripheral Connection Interface* (**PCI**) or Peripheral component interconnect. With the rapid growth of computer evolution, one of the best way you can save money and at same time, to catch-up to modern technology is upgrade your computer and you will need know the slot type to choose the correct matching cards.

Industry Standard Architecture (ISA)

The ISA slot is an 8-bit bus capable of transferring data at the rate of *0.625MB/sec* between the card and the motherboard. Later versions of this bus were 16-bit, capable of **2MB/sec** transfer rate. The ISA is the most widely used bus, because it is the original. They usually appear to be the longer slots on the motherboard.

Enhanced Industry Standard Architecture (EISA)

Since the standard ISA bus appears to be slow the coming of a 32-bit EISA brought major enhancement. Enhanced ISA can now handle demands placed on component in servers or computers that host networks. Also, as discussed under SCSI storage device, the EISA bus is capable of *bus mastering*, which allows components attached to the bus to talk to each other directly without bothering the CPU. Much like SCSI bus mastering absorbs significant overhead from the CPU thus speeding up the computer.

Micro Channel Architecture (MCA)

This is a 32-bit bus, which has several enhanced features, such as bus mastering, identifying devices plugged into it and automatic configuration capabilities over EISA. MCA also produced less electrical interference, reducing errors. In spite of these features however, MCA is no more used in modern computers.

Video Electronics Standard Association (VESA)

This is a very fast interface made up mainly for fast new video cards. Video and graphics data are usually voluminous, so the computer system may require much speed to handle them. This bus provides an interface that can transfer data at **132MB/sec**. To achieve the required high speed the VESA Bus must be connected straight to the CPU's own internal bus, making its location local to the CPU. That's why **VESA Bus** is known as the VESA-Local Bus, or VL-Bus. VESA buses are basically an ISA slot with an extra slot on the end. It measures about 4 inches longer than an ISA slot. Dependency on the CPU is a huge drawback on the VL-Bus, which was limited to the 486 models.

Accelerated Graphics Port (AGP)

Nowadays Pentium computers come with *AGP* slots (*Accelerated Graphics Port)*, which enables the graphics controller to communicate with the computer and access its main memory. The AGP is a high-speed port interface designed to handle 3-D technology, and it stores 3-D textures in the main memory rather than the video memory.

Peripheral Component Interconnect (PCI)

Also called Peripheral Connection Interface, PCI is a more improved version of ISA. It is an Intel-designed bus that features quick communications between a peripheral and the computer's CPU. PCI buses allow additional peripherals to be installed on a system, and they provide plug-and-play capability without any configuring. Each add-on card that slots into this unit contains information about itself that the processor can use to automatically configure the card. It is different than the VL-Bus except that it runs at the same speed. The PCI bus is independent of the CPU. This bus is most popular today with Pentiums, and occasionally present in 486 systems.

Personal Computer Memory Card International Association

(PCMCIA)-This is a special socket in which you can plug removable credit-card size devices. The socket uses a **68-pin** interface to connect to the motherboard or to the system's expansion bus. The circuit cards like memory, hard drives, modems, network adapters, and sound cards, can slot into the PCMCIA socket.

Expansion Slot Standard Sizes

Via PCI, AGP, or ISA interfaces, connections may be established between the external peripherals and the CPU. Depending on how much you pay for, your computer may be loaded with several expansion cards that may allow you to attach many external devices. The slot size is a very important issue, which may guide users when buying PC cards equipment. There are three types of PC cards and as such three types of PC slots, types 1, 2 and 3.

Type 1 Slots: They are usually **3.3**mm thick and hold items such as RAM and flash memory. Type 1 slots are most often seen in palmtop machines or other handheld devices.

Type 2 Slots: are **5**mm thick and I/O capable. These slots are used for I/O devices such as modems and network adapters.

Type 3 Slots: are **10.5**mm thick and used mainly for add-on hard drives. In most cases, Type 3 can handle Type 2 and Type 1.

Motherboard Devices

The motherboard is of particular interest to us since its operation may affect every component in the computer system. The eight common features described above will be further broken down into 63 terms in the list below. Most of these terms will be fully discussed under the chapters where they belong. In the subsequent chapters we will learn about all the common pieces, which can be installed on the motherboard when building a complete and a functional personal computer. The most important piece among them is the CPU, followed by memory, and the input/output devices, which are usually termed as expansion cards and peripheral devices. To better understand these terms the glossary below will assist you to master the most common technical terms used in the personal computer jargon. What is amazing about this glossary is that any term listed in it has its corresponding device for which is worthy to know something about. In other words it represents the summary of computer components.

Component	Descriptions
ACPI	Advanced Configuration and Power Interface: the successor to DPMA for controlling power management and monitoring the health of the system.
ACR	Advanced Communication Riser: a rival riser card architecture to Intel's CNR specification, which emerged at about the same time and offers similar features.
AGP	Accelerated Graphics Port: an Intel-designed 32-bit PC bus architecture introduced in 1997 allowing graphics cards direct access to the system bus (currently up to 100MHz), rather than going through the slower 33MHz PCI bus. AGP uses a combination of frame buffer memory local to the graphics controller, as well as system memory, for graphics data storage, vastly increasing the amount of memory available for 3D textures.
AMR	Audio Modem Riser: an Intel specification that defines a new architecture for the design of motherboards. AMR allows manufacturers to create motherboards without analogue I/O functions. Instead, these functions are placed on a separate card which plugs in perpendicular to the motherboard so that the motherboard and "riser" card form a right angle.
AT Bus	The 16-bit bus started with the IBM-AT (Advanced Technology) systems. It is still the standard interface for most PC expansion cards. It is also known as the ISA (Industry Standard Architecture) bus.
ATA	AT Attachment: the specification, formulated in the 1980s by a consortium of hardware and software manufacturers, that defines the IDE drive interface. AT refers to the IBM PC/AT personal computer and its bus architecture. IDE drives are sometimes referred to as ATA drives or AT bus drives. The newer ATA-2 specification defines the EIDE interface, which improves upon the IDE standard. See also IDE and EIDE.

ATAPI	Advanced Technology Packet Interface: a specification that defines device side characteristics for an IDE connected peripheral, such as CD-ROM or tape drives. ATAPI is essentially an adaptation of the SCSI command set to the IDE interface.
ATX	The predominant motherboard form factor since the mid-1990s. It improves on the previous standard, the Baby AT form factor, by rotating the orientation of the board 90 degrees. This allows for a more efficient design, with disk drive cable connectors nearer to the drive bays and the CPU closer to the power supply and cooling fan.
Baby AT	The form factor used by most PC motherboards in the early 1990s. The original motherboard for the PC-AT measured 12in by 13in. Baby AT motherboards are a little smaller, 8.5in by 11in.
Backside Bus	A special microprocessor bus that connects the CPU to a Level 2 cache. See also Frontside Bus.
BIOS	Basic Input Output System: a set of low-level routines in a computer's ROM that application programs (and operating systems) can use to read characters from the keyboard, output characters to printers, and interact with the hardware in other ways. It also provides the initial instructions for POST (Power On Self-Test) and booting the system files.
Bus Master IDE	Capability of the PIIX element of Triton chipset to effect data transfers from disk to memory with minimum intervention by the CPU, saving its horsepower for other tasks.
Cache	An intermediate storage capacity between the processor and the RAM or disk drive. The most commonly used instructions are held here, allowing for faster processing.
Chipset	A number of integrated circuits designed to perform one or more related functions.
CMOS RAM	Complementary Metal Oxide Semiconductor Random Access Memory: a bank of memory that stores a PC's permanent configuration information, including type identifiers for the drives installed in the PC, and the amount of RAM present. It also maintains the correct date, time and hard drive information for the system.
CNR	Communications and Networking Riser: An Intel riser card architecture that provides expanded audio, modem and networking functions.

Concurrent PCI	An enhancement to the PCI bus architecture that allows PCI and ISA buses to transfer data simultaneously.
DIP Switch	Switch mounted on PC board for configuration options.
DMA	Direct Memory Access: a process by which data moves directly between a disk drive (or other device) and system memory without requiring the involvement of the CPU, thus allowing the system to continue processing other tasks while the new data is being retrieved.
DPMA	Dynamic Power Management System: Intel's extensive set of power management features built in at the chipset level, with particular emphasis on intelligent power conservation and standby facilities.
EIDE	Enhanced Integrated Device Electronics or Enhanced Intelligent Drive Electronics: an enhanced version of the IDE drive interface that expands the maximum disk size from 504Mb to 8.4Gb, more than doubles the maximum data transfer rate, and supports up to four drives per PC (as opposed to two in IDE systems). EIDE's primary competitor is SCSI-2, which also supports large hard disks and high transfer rates.
EISA	Extended Industry Standard Architecture: an open 32-bit extension to the ISA 16-bit bus standard designed by Compaq, AST and other clone makers in response to IBM's proprietary MCA (Micro Channel Architecture) 32-bit bus design. Unlike the Micro Channel, an EISA bus is backward-compatible with 8-bit and 16-bit expansion cards designed for the ISA bus.
ESCD	Region of non-volatile memory used by BIOS and ICU (Intel Configuration Utility) or PnP operating system to record information about the current configuration of the system.
ESDI	Enhanced Small Device Interface: an interface standard developed by a consortium of the leading PC manufacturers for connecting disk drives to PCs. Introduced in the early 1980s, ESDI was two to three times faster than the older ST-506 standard. It has long since been superseded by the IDE, EIDE and SCSI interfaces.
Expansion Bus	An input/output bus typically comprised of a series of slots on the motherboard. Expansion boards are plugged into the bus. ISA, EISA, PCI and VL-Bus are examples of expansion buses used in a PC.

FDD	The interface which allows a floppy or tape drive to be connected to the motherboard.
Frontside Bus	The bus within a microprocessor that connects the CPU with main memory. See also Backside Bus.
Heat Sink	A structure, attached to or part of a semiconductor device that serves the purpose of dissipating heat to the surrounding environment; usually metallic and often aluminium.
Host Adapter	A plug-in board or circuitry on the motherboard that acts as the interface between the system bus and a peripheral device. IDE and SCSI are examples of peripheral interfaces that call their controllers host adapters.
IDE	Integrated Device Electronics or Intelligent Drive Electronics: a drive-interface specification for small to medium-size hard disks (disks with capacities up to 504Mb) in which all the drive's control electronics are part of the drive itself, rather than on a separate adapter connecting the drive to the expansion bus. This high level of integration shortens the signal paths between drives and controllers, permitting higher data transfer rates and simplifying adapter cards. See also EIDE and SCSI.
IRQ	An Interrupt ReQuest signal is generated by a device, to request processing time from the CPU. Each time a keyboard button is pressed or a character is printed to a screen, an IRQ is generated by the requesting device. No two devices can share the same IRQ. A PC has 16 IRQs.
ISA	Industry Standard Architecture: the architectural standard for the IBM XT (8-bit) and the IBM AT (16-bit) bus designs. In ISA systems, an adapter added by plugging the card into one of the 16-bit expansion slots enables expansion devices like network cards, video adapters and modems to send data to and receive data from the PC's CPU and memory 16 bits at a time. See also EISA.
Jumper	Small metal blocks with black plastic handles for enabling or disabling specific functions on a motherboard or expansion card.
Local Bus	A bus which co-exists with the main bus and connects the processor itself to the main memory. PCI is now the standard local bus architecture, having replaced the older VL-Bus.

LPX	A motherboard form factor which allows for smaller cases used in some desktop model PCs. The distinguishing characteristic of LPX is that expansion boards are inserted into a riser that contains several slots and are therefore parallel, rather than perpendicular, to the motherboard.
MCA	Micro Channel Architecture: a 32-bit bus architecture introduced by IBM for their PS/2 series microcomputers. Incompatible with original PC/AT (ISA) architecture.
Motherboard	The PC's main printed circuit board which houses the processor, memory and other components.
NLX	An Intel-designed motherboard form factor. It features a number of improvements over the ATX design providing support for new technologies such as AGP and allows easier access to motherboard components.
Northbridge	Refers to the System Controller component of a Pentium chipset, responsible for integrating the cache and main memory DRAM control functions and for managing the host and PCI buses. See also Southbridge.
PCI	Peripheral Component Interface: the 32-bit bus architecture (64-bit with multiplexing), developed by DEC, IBM, Intel, and others, that is widely used in Pentium-based PCs. A PCI bus provides a high-bandwidth data channel between system board components such as the CPU and devices such as hard disks and video adapters. Superseded the VL-Bus, which was widely used in 486 PCs in the early 1990s.
PIO	Mode Programmed Input Output Mode: a method of transferring data to and from a storage device (hard disk or CD device) controller to memory via the computer's I/O ports, where the CPU plays a pivotal role in managing the throughput. For optimal performance a controller should support the drive's highest PIO mode (usually PIO mode 4).
PIXX	PCI ISA IDE Xcelerator: a key component of the Peripheral Bus Controller chipset, responsible for integrating many common I/O functions found in ISA-based PC systems.
POST	Power-On Self-Test: a set of diagnostic routines that run when a computer is first turned on.
PS/2	An IBM personal computer series introduced in 1987, superseding the original PC line. It introduced the 3.5in floppy disk, VGA graphics and Micro Channel bus. The

	latter has since given way to the PCI bus.
RAS Line	Physical track on motherboard used to select which sides of which SIMMs will be involved in a data transfer. A given chipset supports only a certain number of RAS lines, thereby dictating how many SIMMs can be accommodated. A pair of SIMMs uses one RAS line; a pair of DIMMs uses two.
SCSI	Small Computer System Interface: an American National Standards Institute (ANSI) interface between the computer and peripheral controllers. SCSI excels at handling large hard disks and permits up to eight devices to be connected along a single bus provided by a SCSI connection. The original 1986 SCSI-1 standard is now obsolete and references to "SCSI" generally refer to the "SCSI-2" variant. Also features in Narrow, Wide and UltraWide flavours. See also IDE.
Slot 1	Intel's proprietary CPU interface form factor for Pentium II CPUs. Slot 1 replaces the Socket 7 and Socket 8 form factors used by previous Pentium processors. It is a 242-contact daughter-card slot that accepts a microprocessor packaged as a Single Edge Contact (SEC) cartridge. Communication between the Level 2 cache and CPU is at half the CPU's clock speed.
Slot 2	An enhanced Slot 1, which uses a somewhat wider 330-way connector SEC cartridge that holds up to four processors. The biggest difference from Slot 1 is that the Level 2 runs at full processor speed.
Slot A	AMD's proprietary 242-way connector SEC cartridge used by their original Athlon processor. Physically identical to Slot 1 but electrically incompatible.
Socket 370	Intel's proprietary CPU interface form factor first introduced for its Celeron line of CPUs and subsequently adopted for later versions of the Pentium III family.
Socket 423	Intel's proprietary CPU interface form factor used by its early Pentium 4 processors.
Socket 478	Intel's proprietary CPU interface form factor which replaced Socket 423 with the advent of the 0.13-micron Pentium 4 Northwood core.
Socket 7	The CPU interface form factor for fifth-generation Pentium-class CPU chips from Intel, Cyrix, and AMD.
Socket 8	Intel's proprietary CPU interface form factor used exclusively by their sixth-generation Pentium Pro CPU

	chip. Socket 8 is a 387-pin ZIF socket with connections for the CPU and one or two SRAM dies for the Level 2 cache.
Socket A	AMD's 462-pin CPU interface form factor which replaced Slot A at the time of the introduction of the Thunderbird and Spitfire cores used by AMD's Athlon and Duron desktop processor ranges respectively.
Southbridge	Refers to the Peripheral Bus Controller component of a Pentium chipset, responsible for implementing a PCI-to-ISA bridge function and for managing the ISA bus and all the ports. See also Northbridge.
System Bus	The primary pathway between the CPU, memory and high-speed peripherals to which expansion buses, such as ISA, EISA, PCI and VL-Bus, can connect. Importantly raised from 66MHz to 100MHz in early 1998 with the release of the 440BX Pentium II chipset. Also referred to as the external bus or host bus.
Ultra DMA	A hard drive protocol which doubled the previous maximum I/O throughput to 33 MBps.
USB	Universal Serial Bus: Intel's standard for attaching peripherals to PCs. Designed for low to medium data throughput, it should remove the need to install many devices internally once it gains widespread acceptance.
VESA	Video Electronics Standards Association: the consortium of computer manufacturers responsible for the SVGA video standard and the VL-Bus local-bus architecture.
VLB	VESA Local Bus or VL-Bus: the 32-bit local-bus standard created by the Video Electronics Standards Association (VESA) to provide a fast data connection between CPUs and local-bus devices. The VL-Bus was widely used in 486 PCs, but has since been replaced by the Intel PCI Bus.
VRM	Voltage Regulator Module: used to absorb the voltage difference between a CPU, which may be added in the future and the motherboard.
ZF	Zero Insertion Force: a socket allows a processor to be upgraded easily and without the need for specialist tools. It clamps down on the microprocessor pins using a small lever located to the side of the socket. Socket 5 and Socket 7 are common types of ZIF socket.

Exercise 2.

Students are required to select from the five alternatives labeled A-E the most appropriate answer to each question numbered 1-13

1. The term integrated circuit IC refers to which computer hardware.
   A. Computer machine.
   B. Computer micro-chip
   C. Computer motherboard
   D. Computer microprocessor and motherboard
   E. All of the above

2. What is the difference between a laptop and a desktop computer?
   A. Laptops have more efficient storage capabilities than desktops.
   B. Desktops are upgradeable whereas laptops are not.
   C. Desktops are usually faster than laptops.
   D. Desktop system units are usually separated from the external peripherals, but laptops are not.
   E. Laptops have smarter CPUs than desktop.

3. The term *hardware* refers to:
   A. The computer itself.
   B. All the physical electronic components.
   C. The intelligent part of the computer called the CPU.
   D. Everything within the system unit box only.
   E. All of the above

4. Your Personal Computer is a **system** implies?

    A. It has complex visible part that controls the computer.

    B. The non-visible part, which is grafted to the hardware.

    C. The set of instructions that coordinates all operations.

    D. It contains a collection of chips that work by instruction.

    E. It has interrelated components of hardware and software that work together.

5. Assuming you are to counsel a friend who wants to buy his/her first computer, which of the following computer components will be considered with the highest priority?

    A. Suitable system unit and the input/output peripherals

    B. System unit and the monitor and the printer.

    C. The CPU and the monitor only.

    D. The micro-processor, motherboard, monitor and printer.

    E. The micro-processor, CPU, System Unit and monitor.

6. Which of the following refers to internal peripherals of a standard desktop PC ?

    A. Hard drive   D. Printer

    B. Keyboard   E. All of the above

    C. Mouse

7. The first version of Disk Operation System (DOS) created by Microsoft came to run on Intel chip:

    A. 8088   D. 80286

    B. 8086   E. 80386

    C. Apple II 6502

8. For a typical PC's system unit architecture the basic components commonly include the following.

   A. CPU, Memory and I/O interface.

   B. Microprocessor, I/O Interface.

   C. Control/Datapath, and Memory.

   D. The integrated circuit and memory.

   E. The CPU, Motherboard, and memory.

9. Which of the following is a **major** factor to distinguish between two different CPU models?

   A. speed

   B. storage capacity

   C. price

   D. intelligence

   E. None of the above

10. The internal communications between the CPU and its peripherals occur via which device?

   A. Modems

   B. Ports

   C. Bays

   D. Slots

   E. Bus

11. Which of the following interface devices handles a speed of video display in your PC?

    A. ISA-bus

    B. PCI-bus

    C. AGP-device

    D. Multimedia port

    E. None of the above

12. To connect your PC's system unit to its peripherals for input and output, and other communications, your PC must have?

    A. ports

    B. bays

    C. slots

    D. cards and adapters

    E. modems

13. What does PC motherboard refer to?

    A. The main System unit

    B. The IC or integrated circuit

    C. Circuit board that holds the CPU and its peripherals together

    D. The mini-tower to which all the external peripherals may be attached

    E. Constitutes the main computer hardware.

Computer Hardware - CPU

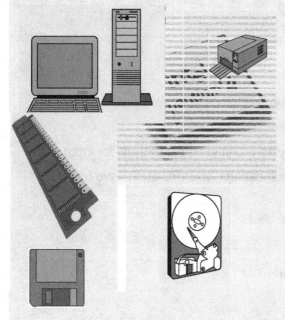

Chapter **3**

o   **The Processor – The Computer Brain**

o   **Structure of CPU**

o   **CPU Clock Cycle And Speed**

o   **Performance of CPU, Multiprocessing**

o   **The Types of CPU**

o   **Principles of Operation- how does it work**

o   **Processor Over-clocking**

The Processor – The Computer Brain

The most important component inside the computer system is the *processor,* which we usually call the **CPU**, or the *central processing unit* (or microprocessor), some people call it a computer *chip*. It is the most active part of the computer. This unit is responsible for all events inside the computer. It controls all internal and external devices, performs arithmetic and logic operations. The CPU is thought of as the brain and the heart of the computer system. The operations a microprocessor performs are called the *instruction set* of this processor. The instruction set is "hard wired" in the CPU and determines the machine language for the CPU. The more complicated the instruction set is, the slower the CPU works.

As shown in *figure 3.1*, the processor is a silicon chip that deciphers and initiates the commands put in the computer system. Each CPU has a measurable performance rated as the *clock cycle* speed of the computer. The speed is measured mostly in *megahertz* or millions of cycles per second, and sometimes in *Gigahertz* (billions of cycles per second), and it is the only means of determining how fast the microprocessor can perform its calculations. The CPU clock cycles will be discussed later in this chapter.

Figure 3.1. An *Intel Pentium 4 Microprocessor and Motherboard*

Microprocessor

Processors differ from one another by the instruction set. If the same program can run on two different computer brands they are said to be **compatible**. Programs written for IBM compatible computers for example will not run on Apple Macintosh computers because these two architectures are not compatible. They have different instruction sets.

There is an exception to this rule. Apple Macintosh with a program like SoftPC loaded can run programs written for IBM compatible PC. Programs like SoftPC make one CPU "pretend" to be another. These programs are called **software emulators**. Although software emulators allow the CPU to run incompatible programs they severely slow down the performance.

Structure of a CPU

To better understand the logic underlying the unique function of the microprocessor, this section will explore the structure of a CPU. A quick look at the Pentium 4 chip in figure 3.2 below will make a CPU appear so simple.

Figure 3.2. An Intel Pentium 4

Intel Pentium 4
2.4 GHz Clock Speed
128-Bit processor
533 MHz Bus Speed

Actually the CPU is more complex than the above figure. The Internal structure of the CPU looks like the illustration in figure 3.3. The CPU's major functional components labeled in figure 3.3 includes the core, cache memory, registers, 64-bits data bus and 32-bits address busses, Integer ALU, branch predictor, execution unit, and floating point unit.

Figure 3.3. Basic Structure of a CPU.

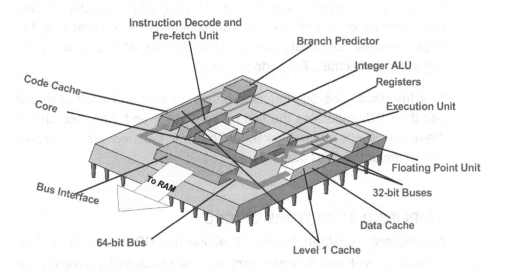

Core: This is the execution cell that combines the Integer ALU, registers, and the main execution unit. The heart of a modern processor is the execution unit. The Pentium has two parallel integer pipelines enabling it to read, interpret, execute and dispatch two instructions simultaneously.

sequence will be executed each time the program contains a conditional jump, so that the Pre-fetch and Decode Unit can get the instructions ready in advance.

Floating Point Unit: This unit is the computing cell for real numbers or decimal fractions. It is the third execution unit in a Pentium, where non-integer calculations are performed.

Level 1 Cache: The cache is a portion of the CPU chip containing a special type of memory. The Pentium has two on-chip caches of 8KB each, one for code and one for data, which are far quicker than the larger external secondary cache. The Pentium IV cache is further discussed under the CPU performance.

Bus Interface: The bus interface brings a mixture of code and data into the CPU, separates the two ready for use, and then recombines them and sends them back out. The bus features are discussed further under the CPU performance.

SIMPLIFIED STRUCTURE OF CPU

As you can see from figure 3.3 above the CPU is composed of several units. I will however simplify this structure by dividing the structural components into three main units *control*, *datapath* and *clock unit*. These basic components constitute the respective brawn and brain of the processor. A simplified version reflecting these units is illustrated in *figure 3.4b* below.

Control - The *control unit* directs and controls the activities of the internal and external devices by sending the signals that determine the operations of the datapath, memory, input and output devices. The control unit also interprets the instructions fetched into the computer, determines what data, if any, are needed, where it is stored, where to store the results of the operation, and sends the control signals to the devices involved in the execution of the instructions.

Figure 3.4. Structure Standard CPU

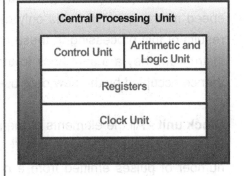

Figure 3.4a. An Intel Pentium III CPU   *Figure 3.4b. A simplified diagram of the CPU*

Datapath - The *datapath* consists of the ***Arithmetic and Logic Unit*** (ALU) and several ***registers*** that work together to perform the arithmetic operations. The CPU processing depends on the datapath. This requires further discussions after learning about the CPU's clock unit.

The arithmetic and logic unit (ALU) is the part where actual computations take place. It consists of circuits, which perform arithmetic operations such as addition, subtraction, multiplication, and division over data received from memory. The ALU is also capable to compare numbers. While performing these operations the ALU takes data from the temporary storage area inside the CPU named **registers**.

Registers - Registers are a group of cells used for memory addressing, data manipulation and processing. Some of the registers are general purpose and some are reserved for certain functions. The CPU registers may include program *counter (PC), instruction decoder, instruction register (IR), process status register, accumulator (ACC)*, and *general purpose registers*. It is a high-speed memory, which holds only data for immediate processing and results of this processing. If these results are not needed for the next instruction, they are sent back to the main memory and registers are thence occupied by the new data used in the next instruction.

Clock unit - All the elements of the processor stay in step by use of a clock, which dictates how fast it operates. The clock rate is the number of pulses emitted from a computer's clock in one second; it determines the rate at which logical or arithmetic operation is performed in a synchronous computer. All activities in the computer system are composed of thousands of individual steps, which should follow in some order in fixed intervals of time. The orderly processing to steps is said to be *pipelining, (see figure 3.5.)* The deeper pipeline stages can account for the creation of today's high-speed processors. The consistent intervals between, these steps of operation are generated by the **clock unit**. Every operation within the CPU takes place at the clock pulse. No operation, regardless of how simple it might be, can be performed in less time than the predefined ticks of this clock. But some operations require more than one clock pulse. The faster the clock runs, the faster the computer performs. The clock rate is measured in *hertz*, ranging from *Megahertz* or million ticks per second to *Gigahertz* (billion ticks per second. Faster systems are created almost yearly. At the moment of this writing the highest clock rate is **3.2GHz**, but by the time you read them you will probably find faster microprocessors already on the market. Larger systems are even faster. Unlike the older systems, most modern microprocessors have their clocks usually incorporated within the CPU. You may read the clock cycle rate below to cover up the missing gap.

PIPELINING FOR A SIMPLE 16-BIT ADDER MACHINE

Figure 3.5. SIMPLE 16-BIT ADDER MACHINE

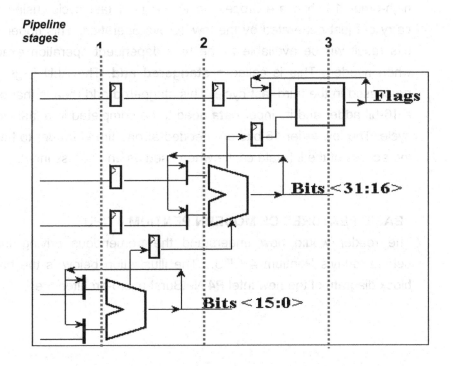

Figure 3.5 above, is a brief illustration of pipelining for the operation of a 16-bit Adder machine. Interested readers may refer to Computer Organization & Design by David A. Patterson, and John L. Hennessey.

The processor does ALU operations with an effective latency of one-half of a clock cycle. It does this operation in a sequence of three fast clock cycles (the fast clock runs at twice the main clock rate) as shown in Figure 3.5, above. In the first fast clock cycle, the low order 16-bits are computed and are immediately available to feed the low 16-bits of a dependent operation in the very next fast clock cycle. The high-order 16- bits are processed in the next fast cycle, using the carry out just generated by the low 16-bits operation. This upper 16-bits result will be available to the next dependent operation exactly when needed. This is called a **staggered add**. The ALU flags are processed in the third fast cycle. This staggered add means that only a 16-bit adder and its input *data* need to be completed in a fast clock cycle. The low order 16 bits are needed at one time in order to begin the access of the L1 data cache when used as an address input.

BASIC FEATURES OF MODERN PENTIUM 4 CPU

The reader would now understand the tremendous driving force behind today's Pentium 4 CPU. The illustration below is the basic block diagram of the new Intel **P4** NetBurst micro-architecture.

Figure 3.6. Basic block diagram of Intel Pentium IV

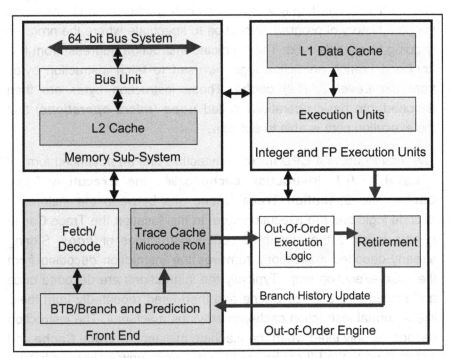

As you can see in the figure 3.6 above, there are four main sections: the **in-order front end**, the **out-of-order execution engine**, the **integer and floating-point execution units**, and **the memory subsystem**.

In-Order Front End

The in-order front end is the part of the machine that fetches the instructions to be executed next in the program and prepares them to be used later in the machine pipeline. Its job is to supply a high-bandwidth stream of decoded instructions to the out-of-order execution core, which will do the actual completion of the instructions. The front end has highly accurate branch prediction logic that uses the past history of program execution to speculate where the program is going to execute next. The predicted instruction address, from this front-end branch prediction logic, is used to fetch instruction bytes from the *Level 2, (L2) cache*. These instruction bytes are then decoded into basic operations called *uops* (*micro-operations*) that the execution core is able to execute.

The Intel Pentium 4 CPU micro-architecture has an advanced form of a **Level 1, (L1) instruction cache** called the **Execution Trace Cache**. The **Execution Trace Cache** sits between the instruction decode logic and the execution core. In this location the Trace Cache is able to store the already decoded instructions or *uops.* Storing already decoded instructions removes the instruction decoding from the main execution loop. Typically the instructions are decoded once and placed in the Trace Cache and then used repeatedly from there like a normal instruction cache on previous machines. The instruction decoder is only used when the machine misses the **Trace Cache** and needs to go to the **L2 cache** to get and decode new instruction bytes.

Out-of-Order Execution Logic

The **out-of-order execution engine** is where the instructions are prepared for execution. The out-of-order execution logic has several buffers that it uses to smooth and re-order the flow of instructions to optimize performance as they go down the *pipeline* and get scheduled for execution. Instructions are aggressively re-ordered to allow them to execute as quickly as their input operands are ready. This out-of-order execution allows instructions in the program following delayed instructions to proceed around them as long as they do not depend on those delayed instructions. Out-of-order execution allows the execution resources such as the **ALUs** and the cache to be kept as busy as possible executing independent instructions that are ready to execute.

The **retirement logic** is what reorders the instructions, executed in an out-of-order manner, back to the original program order. This retirement logic receives the completion status of the executed instructions from the execution units and processes the results so that the proper architectural state is committed (or retired) according to the program order. The Pentium 4 processor can retire up to three **uops** per clock cycle. This retirement logic ensures that exceptions occur only if the operation causing the exception is the oldest, non-retired operation in the machine. This logic also reports **branch history** information to the branch predictors at the front end of the machine so they can train with the latest known-good branch-history information.

Integer and Floating-Point Execution Units

The execution units are where the instructions are actually executed. This section includes the register files that store the integer and floating-point data operand values that the instructions need to execute. The execution units include several types of integer and floating-point execution units that compute the results and also the **L1 data cache** that is used for most load and store operations.

Memory Subsystem
Figure 3.6 also shows the memory subsystem. This includes the **L2 cache** and the system bus. The **L2 cache** stores both instructions and data that cannot fit in the **Execution Trace Cache** and the L1 data cache. The external system bus is connected to the backside of the second-level cache and is used to access main memory when the L2 cache has **a cache miss**, and to access the system I/O resources.

The CPU Processing Power - Internal Data communication

The communications between the CPU and the peripherals (surroundings) occur through the connection device called the bus. Computers may have an *Industry Standard Architecture* **(ISA)** bus which is 16-bits or a *Peripheral Component Interconnect* **(PCI)** bus, which is mostly 32-bits. This type of bus was labeled as 64-bits bus in figure 3.6 above. The PCI bus is twice as fast as the ISA models. The two main features of the CPU are:

1. CPU speed measured in millions of instructions per second; Pentium IV CPU speed is now measured in billions of instructions per second. That's in Gigahertz, (GHz). 1GHz is equivalent to 1 billion instructions per second.
2. The word size, meaning the number of bits that can travel in the datapath simultaneously. It is further described as the number of bits the CPU can process at a time.

At this point *datapath* is of a particular interest to us when it comes to determining the power of the computer machine. It refers to how many bits of information can travel in parallel at one time. For example, a *16-bit datapath* is a **16-wire path** on which 16 bits of information can travel at any time. This results in roughly 2 bytes of information processed at a time. Therefore, such a CPU is said to be a **16-bit CPU**. Today's CPUs can handle much more traffic. We have 32-bit and 64-bit processors, although only a few expansion cards can take advantage of this. Similarly, an *8-bit* processor can handle *8 bits* of data simultaneously. Let's take for example that you want to add up two four-digit numbers like **3591** and **1325**, using three different CPU's. For an *8-bit* CPU, it will take *four operations*, a separate operation for addition of each of the four digits. A *16-bit* processor will be able to work on two digits simultaneously, so it will need only *two* *operations*. A *32-bit* processor will complete addition in *one operation.* Remember, that how fast data is processed will depend on the clock cycle rate of the CPU.

Intel 80486 is a 32-bit processor so it can manipulate twice as much data at one time as say Intel 80286, which is a 16-bit processor. As you can see for yourself in table 3.3 below, the maximum possible word size at the time of this writing is 64bits. This word size comes with Pentium IV operating at 2.4 GHz clock cycle speed. Most supercomputers have such powerful processors. What is amazing is, that modern technology allows us to produce 64-bit processors for personal computers. The 32-bit processors are installed in many personal computers, minicomputers, and in most mainframes.

CPU Address Bus

The address bus is different from the above data bus. The address bus is the wire path that carries addressing information that points to where in the computer memory to look for certain data. Each wire carries one bit of information. Since computers think in binary (a number system based on **0's** and **1's**), a **2-bit** address bus would provide four addresses (00, 01, 10, 11). A 3-bit bus would provide 8 addresses (000, 001, 010, 011, 100,101, 110, and 111), and so on. The number of address will equal 2^x for **x-bits** address bus. The point of all this is that the address bus built into the CPU dictates how much memory the computer can have. A **286** computer has a *24-bit* address bus, thus providing *16,777,216* addresses. This means the 286 CPU can only address 16 MB of memory. The 386 and 486 both have 32-bit address buses, which provide 4,294,967,296 addresses.

These processors can address 4,096 MB of memory, even though no motherboards have enough slots to hold this much memory. To make these addresses available for use, the operating system must implement a computing concept known as *virtual memory*, which will be discussed further under storage memories in the next chapter.

Computers, like other machines, follow the *I-P-O* process which implies *Input-Process-Output.* To process raw data there must be an input, which goes through processing time and finally sent to the output when processing is complete. The CPU controls the peripheral devices, thus making it possible to coordinate all these events at the same time.

o   Input is data entered into your computer from an input unit, such as a keyboard, mouse or a scanner.

o   Process is computing and manipulating data by the "intelligence" of the computer machine. This is the main job of the CPU.

o   Output is data processed into meaningful information that can appear in an output unit, such as a monitor, or a printer.

Remember that data is a raw material while information is data processed into meaningful form as an output or meaningful data.

The CPU's Speed

The speed of your PC is the rate at which the CPU can process a given information. It is also known as clock speed of the CPU. The speed of your CPU directly influences the speed of your entire computer system. There are however, other components such as disk drive speed, memory space, network speed, and video card usage that may also affect the speed of your PC. The CPU speed is an important feature to consider when deciding to buy a computer. Other factors that may affect the CPU speed are discussed further down this chapter under CPU performance.

The CPU speed is measured mostly in *hertz* (Hz), that is, *number of operations (cycles) that can be done per second*. Currently, we usually use **MHz** or Megahertz (millions of cycles per second) and sometimes, in **GHz** or Gigahertz (billions of cycles per second) because today's CPUs are too fast to measure in Hz. The unit hertz is the number of instruction cycles the CPU can perform in a single second. The latest CPUs in personal computers have an average clock speed of 400 Megahertz (MHz) to 2.6 Gigahertz (GHz). Currently, computer CPU running at any speed below 400MHz is considered outdated. As of now the latest personal computer CPU chip on the market is **Pentium IV** with 2.0 to 3.2+ **GHz** plus execution speed that is a higher level performance than the Pentium I, II and III. Processor clock speed is measured these days in GigaHertz (GHz).

CPU Clock Cycle

To avoid confusion, and impediments for newcomers in the industry, modern computers are designed based on abstraction where several details are hidden. This sub-topic may uncover just a little bit of CPU instruction cycles or *CPI,* which stands for *cycles per instruction*. Modern Computers may process data using four basic types of instruction sets including *arithmetic*, *data transfer*, *conditional branch*, and *jump*, each of which can be sub-categorized. Underneath the scene, each basic instruction has been assigned an operation code or *task code,* which in turn corresponds to the estimated number of clock cycles needed to complete a task. A task code has an assumed number of cycles needed to accomplish that particular task. In table 3.1 below, we assume that each of the four basic instruction types carries an estimated average clock cycle values or, average CPI for the instructions. An application of this instruction table is provided in *example 3.1* below

Table 3.1.The four basic computer instructions and their assumed average clock cycles.

Instruction Types	Average Clock Cycles
Arithmetic	1.0 clock cycles
Data Transfer	1.4 clock cycles
Conditional branch	1.7 clock cycles
Jump	1.2 clock cycles

Example 3.1

Given the arithmetic operation   **A = B + C**;

a) How many clock cycles may be involved in this operation?

b) How long will it take your Pentium III running at CPU speed of 850 MHz to process this information?

The Solution:

(*Note: This solution assumes that B and C have values less than 64-bits each*)

A) The breakdown of the instruction **A = B + C** will be in the following sequence of operations. (*converted to low level assembly language*):

1. Load **B** from memory to **CPU Register B**... **1.4 cycles**
2. Load **C** from memory to **CPU Register C**... **1.4 cycles**
3. Add Registers (**B + C**).................. **1.0 cycles**
4. Store results to memory block **A** .... **1.4 cycles**
5. Total cycles needed for **Addition**....... **5.2 cycles**

We would need approximately **5 clock cycles** to complete the instruction **A = B + C**

B) The Pentium III has 850 MHz speed.

⇔ 850×10^6 Hertz

⇔ 850×10^6 clock cycles per second

⇔ $\dfrac{850 \times 10^6 \text{ clock cycles}}{1 \text{ second}}$

Therefore seconds in 5 clock cycles

⇔ $(5 \text{ clock cycles}) \bigg/ \left(\dfrac{850 \times 10^6 \text{ clock cycles}}{1 \text{ second}} \right)$

⇔ $\dfrac{(5 \text{ clock cycles}) \times 1 \text{ second}}{(850 \times 10^6 \text{ clock cycles}} = \dfrac{0.00588}{10^6}$ seconds

⇔ $\dfrac{0.00588 \times 10^9 \text{ nanoseconds}}{10^6}$

⇔ **5.9 nanoseconds**

Performance Of Modern CPU

When upgrading or buying a new PC - the first question that usually comes in mind is how much faster can my computer run? And as was discussed above when most people think of PC speed, they think of **processor speed**. Depending on how you use your PC other components such as the memory, video card, or hard-drives can greatly influence PC performance. But the biggest driver behind performance has and probably always will be the speed of the central processing unit (CPU).

Because of this, the processor is still arguably the most important component to consider when upgrading or buying a new computer. Processors are sold with their clock speed being the headline, which is measured these days in **GigaHertz** (GHz). A **1GHz** CPU speed will cycle **1 billion** instructions per second. The higher the number of **GHz**, the faster the clock speed runs at, hence more instructions per second can be processed, and so, assuming there are no significant impacts caused by other components in your PC, the faster your PC can run. There are, however, a number of other processor features that can affect the CPU performance. The most interesting features that can greatly influence the speed of a CPU are listed below:

- Cache size
- Process (Mfg. Tech = manufacturers technology)
- Front-Side Bus Speed
- Socket package Type

The table 3.2 below shows a short list of significant features to consider when deciding on which CPU will produce the best performance. It is important to note that the above list is not exhaustive. The CPU giants like Intel Corporation and AMD keep on producing higher performance CPU year after year. What matters most at this point is the ability to choose a CPU that carries the best combination of these features described above.

Table 3.2. Listing of The Most Significant Features of Pentium IV CPU

Spec#	CPU Speed	Bus Speed	Mfg. Tech	Stepping	Cache Size	Package Type
SL7J9	3.60 GHz	800 MHz	90 nm	D0	1 MB	775 pin PLGA
SL7KN	3.60 GHz	800 MHz	90 nm	D0	1 MB	775 pin PLGA
SL7KM	3.40 GHz	800 MHz	90 nm	D0	1 MB	775 pin PLGA
SL7J8	3.40 GHz	800 MHz	90 nm	D0	1 MB	775 pin PLGA
SL793	3.40 GHz	800 MHz	0.13 micron	D1	512 KB	478 pin PPGA
SL7KL	3.20 GHz	800 MHz	90 nm	D0	1 MB	775 pin PLGA
SL7J7	3.20 GHz	800 MHz	90 nm	D0	1 MB	775 pin PLGA
SL792	3.20 GHz	800 MHz	0.13 micron	D1	512 KB	478 pin PPGA
SL6WE	3.20 GHz	800 MHz	0.13 micron	D1	512 KB	478 pin PPGA
SL6JJ	3.06 GHz	533 MHz	0.13 micron	C1	512 KB	478 pin PPGA FC-PGA2
SL6S5	3.06 GHz	533 MHz	0.13 micron	C1	512 KB	478 pin PPGA FC-PGA2
SL6WK	3 GHz	800 MHz	0.13 micron	D1	512 KB	478 pin PPGA FC-PGA2
SL79L	3E GHz	800 MHz	90 nm	C0	1 MB	478 pin PPGA
SL7J6	3.00 GHz	800 MHz	90 nm	D0	1 MB	775 pin PLGA
SL78Z	3 GHz	800 MHz	0.13 micron	D1	512 KB	478 pin PPGA
SL6WU	3 GHz	800 MHz	0.13 micron	D1	512 KB	478 pin PPGA
3L	2.80 GHz	533 MHz	0.13 micron	C1	512 KB	478 pin PPGA FC-PGA2
SL7J5	2.80 GHz	800 MHz	90 nm	D0	1 MB	775 pin PLGA
SL7K9	2.80A GHz	533 MHz	90 nm	D0	1 MB	478 pin PPGA
SL6K6	2.80 GHz	533 MHz	0.13 micron	C1	512 KB	478 pin PPGA FC-PGA2
SL6PE	2.66 GHz	533 MHz	0.13 micron	D1	512 KB	478 pin PPGA FC-PGA2

The above table shows a sample list of Pentium IV CPUs with different clock speeds and associated features. The listed features may account for the ultimate driving force of today's CPU high speed. Altering any of the features will directly influence the clock speed of the CPU.

Processor Cache Size

The most significant of these features are the **processor cache sizes.** A cache is a portion of the CPU chip containing a special type of memory. It is called **cache** because it's a temporary "hidden" storage space on the actual CPU chip, where repeatedly used instructions are placed and called-back from. Such a system dramatically improves the overall CPU speed, as the CPU references these common instructions repeatedly, by dipping into the cache local to the CPU, instead of having to go to all the trouble of asking for it from the main-memory via the motherboard.

A CPU has up to *three levels* of cache, designated **L1, L2** and **L3.** The **L1 Cache** is the portion of memory closest to the CPU, where a small number of the most commonly used instructions are kept for re-use, and is the fastest and smallest cache, typically of a size between 2KB and 16KB.

The **L2 cache** is the next level of cache, and varies between 128KB and 2MB. **L2** is the most important in terms of buyer's choice, since CPU manufacturers vary this quantity significantly to differentiate different versions of a CPU having the same clock speed. A CPU having a 1MB cache will usually operate much faster than the same CPU having only a 128KB cache. Because of the additional number of memory transistors required on a large cache CPU, a higher price tag will result.

The **L3 cache** is the next level of cache after L2, and perhaps the highest level of the processor sub-memory system. Level 3 cache comes mostly in sizes of **2MB**, and it's a CPU processing speed booster, that makes a great difference under the **extreme-loading** such as 3D games.

Silicon Process Technology

The **process** or *manufacturer's technology* is another significant influence on chip buying decisions, which refers to the way in which the chip was manufactured, and specifically refers to the standard transistor size used. **Process** is significant because size is important in CPU micro-architecture - the **smaller** the better. Electrons have less distance to travel and can arrive earlier, so higher clock speeds are possible. The smaller transistors also use less power, which is another reason why higher clock speeds are available, since a limitation on clock speed has long been the problem of how to dissipate the heat generated from the power usage.

If a processor is available in two different processes for example 90nm and 130nm, one would think, based on the above, that the smaller process CPU would be faster. In theory this is the case, but in practice, the two are often comparable in performance, usually because the chip manufacturer, in introducing the brand-new process, has not quite optimized it in their haste to launch it onto the highly competitive chip market. The performance potential on the smaller process, however, is significantly greater, since it is at the beginning of a new generation of chips. This can have more of an influence on motherboard selection than on CPU purchase, as a motherboard, which supports a new process will be highly future-compatible, compared to one that will only support the old process.

For instance, the Intel Pentium 4 **"Prescott"** CPU uses **90nm** process - that's **90 nano-meters** or about **1/100,000th** of a millimeter. In other words, pretty small. Process has been reducing in size year on year - the Prescott CPU process is ten times smaller than that of the original Pentium when it was introduced in the mid 1990s.

The last type of Pentium 4 Socket 478 variants is the **Extreme-Edition**. This very expensive processor comes with a 2MB on-chip L3 cache, which makes a difference under **extreme loading**, such as 3D games.

Bus Speed

The **Bus Speed,** often referred to as the **Front –Side-Bus Speed (FSB)** is a very significant feature affecting the processors speed.

For example the Pentium IV processor has a system bus with 3.2 Gigabytes per second of bandwidth. This high bandwidth is a key enabler for applications that stream data from memory. This bandwidth is achieved with a 64-bit wide bus capable of transferring data at a rate of 400 MHz. It uses a source-synchronous protocol that quad-pumps the 100 MHz bus to give 400 million data transfers per second. It has a split-transaction, deeply pipelined protocol to allow the memory subsystem to overlap many simultaneous requests to actually deliver high memory bandwidths in a real system. The bus protocol has a 64-byte access length.

Package Type

Two different types of Pentium IV Processor are currently available in the Socket 478 package - **Northwood** and **Prescott.** The newer variant is Prescott, and the main difference between the two is that Prescott processors have a 90nm process instead of 130nm. The Prescott variant also comes with a 1MB L2 cache instead of the 512KB cache on the Northwood. The last type of Pentium IV Socket 478 variants is the Extreme-Edition. This very expensive processor comes with a 2MB on-chip L3 cache, which makes a difference under extreme loading, such as 3D games. The Pentium IV's are also said to be Hyper-Threading (HT) enabled.

The Prescott variants are the next sub-generation of Pentium IV processors, and will, in time, likely take the Pentium 4 processor up to 4GHz with Socket package of 775 "**land-grid-array**" or LGA755. All **LGA775** Pentium IV processors use the new Prescott 90nm process and will only work with new LGA775 motherboards, which have the special 775-connector land-grid-array to receive this pin-less processor package.

Monitoring The Processor Performance

We now know that the speed of the CPU is a very important factor when deciding which computer to buy. The CPU speed has a direct impact on the performance of the computer system unit. The following factors can also be used to monitor and optimize the performance of your CPU.

- o Percentage (%) Processor Time
- o Interrupt per second
- o Processor Queue Length

Percentage (%) Processor Time: The percentage of time the processor is busy performing tasks. Computer systems experts believe the desired value would preferably, be below **80%.** Evidently, if the percentage of time the processor is busy performing tasks would have been **100%**, newly user commands would never get the chance to be executed within a reasonable time. Interrupt responses be greatly affect. The computer would be thinking almost for ever.

Interrupt per second: This is the number of device interrupts the processor is handling each second. Desired value is below *3500* on a Pentium CISC or RISC computer. (CISC and RISC are discussed further down this chapter.)

Processor Queue Length: The number of outstanding requests the processor has in the queue. Desired value is under **two (2)**. The percentage processor time may directly impact this value. The higher the % processor time, the higher the number of the processor's outstanding interrupt request will remain waiting in the queue.

You can use programs such as Windows NT/2000 Performance Monitor to monitor the performance of CPU.

Figure 3.7. Example of Windows Task Manager display current processes

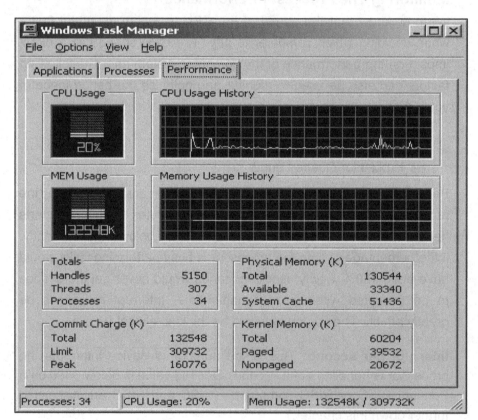

The figure 3.7 above displays the content of windows task manager of Intel Pentium III 850 MHz CPU. You can see in this particular example, the processor was handling 34 processes simultaneously, and the percentage (%) processor time is only 20%.

NB: When you find a CPU is not performing well, do not buy another CPU immediately. The problem may be caused by insufficient, RAM, Hard Disk Space, bad Program design or bad Peripherals. You need to identify the problem first.

Multiprocessing And Multitasking

To boost performance, some PC users buy multiprocessing computers. **Multiprocessing** is the use of two or more independent CPUs linked together. In this environment more than one instruction, can be executed by the CPU at the same time. With multiprocessing, you can share loading among CPUs, gaining speed and saving money. Many server and industrial computers use multiprocessing to increase their performances. To implement multiprocessing, you need:

o   A motherboard, which can hold two or more CPUs.
o   An Operating System that can take the advantage of multiprocessing.

On the other hand one processor can perform several tasks at the same time, and in this case we call it **multi-tasking**. Operating systems like Microsoft windows (NT, 95/98, and 2000) support multi-tasking functions. The figure 3.8 below displays the content of windows task manager. You can see in this particular example, the processor was handling 34 processes simultaneously.

Figure 3.8. Example of Windows Task Manager display current processes

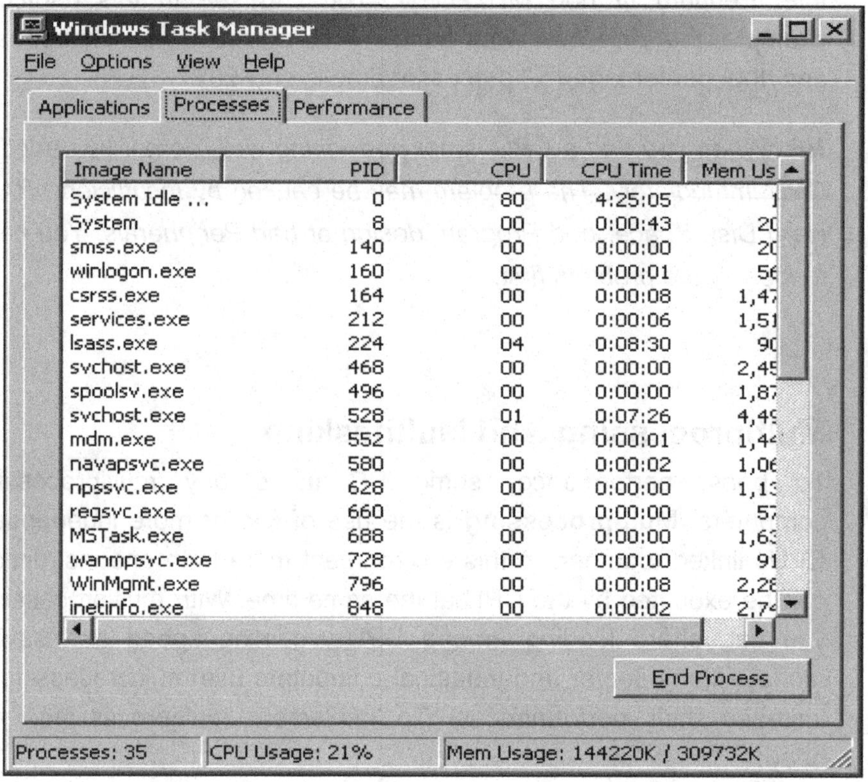

Image Name	PID	CPU	CPU Time	Mem Us
System Idle ...	0	80	4:25:05	1
System	8	00	0:00:43	20
smss.exe	140	00	0:00:00	20
winlogon.exe	160	00	0:00:01	56
csrss.exe	164	00	0:00:08	1,47
services.exe	212	00	0:00:06	1,51
lsass.exe	224	04	0:08:30	90
svchost.exe	468	00	0:00:00	2,45
spoolsv.exe	496	00	0:00:00	1,87
svchost.exe	528	01	0:07:26	4,49
mdm.exe	552	00	0:00:01	1,44
navapsvc.exe	580	00	0:00:02	1,06
npssvc.exe	628	00	0:00:00	1,13
regsvc.exe	660	00	0:00:00	57
MSTask.exe	688	00	0:00:00	1,63
wanmpsvc.exe	728	00	0:00:00	91
WinMgmt.exe	796	00	0:00:08	2,28
inetinfo.exe	848	00	0:00:02	2,74

End Process

Processes: 35   CPU Usage: 21%   Mem Usage: 144220K / 309732K

HOW DOES THE CPU PROCESS DATA?

The bottom line most readers are really itching to know about computers is the mystery behind how CPUs process data. The transistors process information in bits corresponding to binary digits. Matching a set of bits to character codes, there are two popular standards namely **ASCII** and **EBCDIC**. ASCII stands for *American Standard Committee for Information Interchange*, and EBCDIC stands for *Extended Binary Coded Decimal Interchange Code.* ASCII and EBCDIC will be fully discussed under character sets. While the ASCII standard matches a set of 7-bits to process a meaningful character, the EBCDIC matches 8-bits. For example let's assume your CPU was built on ASCII standard, and you typed the letter "A" on your standard keyboard. The character "A" can be processed in your PC by following the steps below, before it will appear on your output device; that is either on your screen or on your printer.

1. *The keyboard will send electric signals corresponding to the letter "**A**" to the CPU's temporary memory.*
2. *Electric signals:* **ON** OFF OFF OFF OFF OFF **ON** ⟹ "A"
3. *The switches* **ON** OFF OFF OFF OFF OFF **ON** ⟹ 1000001_2
4. *The binary digits:* **1000001_2** corresponds ⟹ 65
5. *In ASCII, the character code* **65** *matches the letter* "A"
6. *Thus output is the letter "**A**" on the screen or on printer*

How does the system display the letter "A" is another story discussed under video cards in chapter 6.

How Does THE CPU Understand Binary Language - 0,1?

The hardware is everything you saw in figure 2.1 plus all the internal devices that are covered in the system unit box. That is a combination of the mini-tower (i.e. CPU block), the monitor, the keyboard and the mouse. Also as you can see from *figure 2.1* in chapter two, it is also the solid electronic part of the computer. A modern digital computer has a hardware part that is largely a collection of electronic switches. These switches are logic gates wired in a way to represent and control the routing of data elements that are ***binary digits*** called ***bits***.

Logic gates are in fact, digital switches. In binary numerals only two digits **0**, and **1**, can be used for computing. Inside the computer circuit board the **ON** *switch* equals the digit **1** and the **OFF** *switch* equals the digit **0**. This explains why **0** and **1** are the only two letters your PC can understand. The principal explanation behind this will probably require a separate book. The next topic will outline what occurs in the switch.

Principles Of Operation

Basically all computer processors operate under the same principles. They all take signals in the form of **0**s and **1**s (***data input***), manipulate them according to a well defined instruction set, and produce an output in the form of **0**s and **1**s. The voltage on the line at the time a signal is sent determines whether the signal is a **0** or a **1**. On a 3.3-volt system, an application of 3.3 volts means that it's a **1**, while an application of **0** volts means it's a **0**.

Processors work by reacting to an input of **0**s and **1**s in specific ways dictated by instruction set, and then return the processed bits or an output, based on the decision. The decision itself happens in a circuit called a ***logic gate***, each of which requires at least one ***transistor***, with the inputs and outputs arranged differently by different operations. Modern processors contain millions of transistors making the logic system look more complex. The processors logic gates work together to make decisions using ***Boolean logic***, which is based on the algebraic system established by mathematician called ***George Boole***. The main Boolean operators are **AND, OR, NOT**, and **NAND** (not AND); many combinations of these are possible as well. The table below summarizes the fundamentals of Boolean algebraic system used in processors.

Table 3.2. A Truth table computing Boolean values

Input A	Input B	A AND B	A OR B	NOT A	NOT B	A NAND B
1	1	1	1	0	0	0
1	0	0	1	0	1	1
0	1	0	1	1	0	1
0	0	0	0	1	1	0

An **AND** gate outputs a **1** only *if both its inputs were* **1**s. An **OR** gate outputs a **1** *if at least one of the inputs was a* **1**. And a **NOT** gate takes a single input and reverses it, outputting 1 if the input was 0 and vice versa. **NAND** gates are very popular, because they use only **two** transistors instead of the **three** in an AND gate, yet provide just as much functionality.

A quick look at the simple **AND** and **OR** logic-gate circuits below, shows how the circuitry works. Each of these gates acts on two incoming signals to produce one outgoing signal. *Logical AND* means that both inputs must be **1** in order for the output to be **1**; *logical OR* means that either input can be **1** to get a result of **1**. In the **AND** gate, both incoming signals must be high-voltage (or a logical 1) for the gate to pass current through itself.

Figure 3. 9. Basic Logic Gates For Logical Operators AND and OR

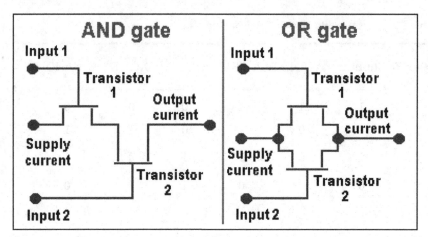

In addition, the processor uses gates in combination to perform arithmetic functions; it can also use them to trigger the storage of data in memory.

Logic gates – As already mentioned, logic gates operate via hardware known as a switch. In the past, the switches were actually physical switches, but today the flow of current is automated. The most common type of switch in computers now is a transistor known as a **MOSFET,** which stands for metal-oxide semiconductor field-effect transistor. This kind of transistor performs a simple but crucial function: When voltage is applied to it, it reacts by turning the circuit either on or off. Most PC microprocessors today operate at **3.3V**, but earlier versions including some Pentiums operated at **5V**. With one commonly used type of MOSFET an incoming current at or near the high end of the voltage range switches the circuit on, while an incoming current near **0** switches the circuit **off**.

In figure 3.10, the line graph below shows the increased trend of transistors used in the manufacturing of microprocessors. This graph was obtained from the Intel website. The calibrations on the vertical axis represent the number of transistors in multiples of thousands or Kilos (K). The Horizontal axis indicates the years in which the microprocessors version was released. For example in 1989 Intel Processor 80486 DX was released. And it carried about 1.6 million transistors, whilst Pentium II released in 1997 loaded 7.5 million transistors. Remember that by the time this book reaches the market, another generation will probably be out. You have to watch out for newer versions.

Figure 3.10. Number Of Transistor Usage Per CPU Generation From 1970-2002

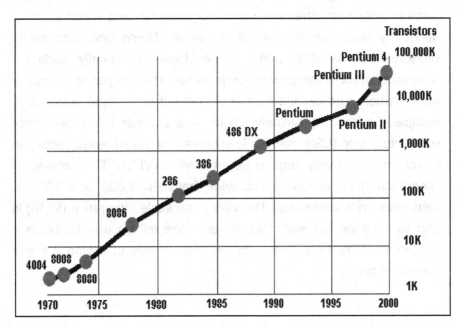

CPU AS INTEGRATED CIRCUIT (IC) LSI, VLSI

Nowadays your CPU contains several millions of **MOSFETs** that work together, to react to the instructions from a program, to control the flow of electricity through the logic gates to produce the required result. Again, each logic gate contains one or more transistors, and each transistor must control the current so that the circuit itself will perform one of these operation at a given time:

1. switch from **off** to **on**,

2. switch from **on** to **off**, or

3. stay in its current state.

The flow of electricity through each gate is controlled by that gate's transistor. However, these transistors are not individual and discrete units. Instead, large numbers of them are manufactured from a single piece of silicon (or other semiconductor material) and linked together without wires or other external materials. These units are called integrated circuits (**ICs**), and their development basically made the complexity of the **microprocessor** possible. The integration of circuits went further, just as the first ICs connected multiple transistors, multiple ICs became similarly linked, in a process known as large-scale integration (**LSI**); eventually such sets of ICs were connected, in a process called very large-scale integration (**VLSI**). The large-scale integration (LSI) is said to combine between 3,000 and 100,000 transistors on a single chip. The very large scale integration (**VLSI**) is said to combine between 100,000 and one million transistors on a single chip. Note here that a chip is another way of saying a single integrated circuit.

Types Of Chips - CISC And RISC Architectures

Whether it is LSI or VLSI two types of microprocessors have dominated the PC industry - Intel's Pentium and Motorola's PowerPC. The two families of CPUs also standout to be the top two competing CPU architectures of the last two decades. The Intel architecture is being classed as a CISC chip, Motorola's version is known as a RISC chip. You can recall in chapter one that the Instruction sets define the basis for CPU architecture. Evidently, the major difference underlying the CISC and RISC architectures rest on their instruction sets.

CISC - CISC means *complex instruction set computer*. This Intel breed is the traditional architecture of a computer, in which the CPU uses micro-code to execute a very comprehensive instruction set. These may be variable in length and may use all addressing modes, requiring complex circuitry to decode them.

RISC - RISC stands for *reduced instruction set computer*. In RISC architecture the CPU keeps instruction size constant, bans the indirect addressing mode and retains only those instructions that can be overlapped and made to execute in one machine cycle or less. Producing architectures like RISC will obviously have its pros and cons. One advantage of RISC CPUs is that they can execute their instructions very fast because the instructions are so simple. Another, perhaps more important advantage, is that RISC chips require fewer transistors, which makes them cheaper to design and produce. RISC machines are both cheaper and faster, and are therefore the machines of the future.

The major disadvantage of RISC chips rest on the compilers having to generate software routines to perform the complex instructions that are performed in hardware by CISC computers. Researchers argue that this may not worth the trouble because conventional microprocessors are becoming increasingly fast and cheap anyway.

Today CISC and RISC implementations are becoming more and more alike. Many of today's RISC chips support as many instructions as yesterday's CISC chips and, conversely, today's CISC chips use many techniques formerly associated with RISC chips. Even though the CISC champion, Intel, used RISC techniques in its 486 chip and has done so increasingly in its Pentium family of processors. Also, today, other microprocessors are compatible with these two leading microprocessors - Intel and Motorola. For example, AMD-K5 is compatible with Pentium.

Evolution of CPUs (Processor chip)

In the last page of chapter one you read about the evolution of computers. You have learned in this chapter that the CPU is the actual computer. It has now become obvious to tie the most telling factor of your computer speed to how fast the microprocessor can process data. In other words, the faster your computer CPU can process data, the more quickly your PC can do its job. Modern computer CPUs may be equipped with a microprocessor that comes in various models. Each model has a name that distinguishes one model from the other. The model names are originals given by the manufacturers like Intel Corporation. You may recall in chapter one that Intel Corporation designed the microprocessor model 8088/8086, which gave birth to our modern Personal Computers. Example of CPU models may range from 8088 to 80-586 or Pentium of Intel chips or processors.

How did it happen?

A Computer chip is just an aggregation of the **OR** and **AND** logic gates. The CPU is another name for the computer chip or microprocessors. Modern day microprocessors contain tens of millions of microscopic transistors that are used in combination with resistors, capacitors and diodes, to make up logic gates. Several Logic gates combine to make up integrated circuits(ICs), and ICs constitute the electronic systems. To do just that the Intel Corporation is well known as the world's most famous processor manufacturing giant, because this company was the first to create high-level integration of all the processor's logic gates into a single complex processor called **chip**.

CPU Generations- First to Eighth

The first chip was the Intel 4004 - released in late 1971. This was a 4-bit microprocessor, intended for use in a calculator. It processed data in 4bits, but its instructions were 8bits long. Program and data memory were separate, 1KB and 4KB respectively. There were also sixteen 4-bit (or eight 8-bit) general-purpose registers. The 4004 had 46 instructions, using only 2,300 transistors in a 16-pin DIP and ran at a clock rate of 740kHz (eight clock cycles per CPU cycle of 10.8 microseconds). **DIP** stands for a Dual In-Line Package chip housing with pins on each edge. In fact along with gradual progress significant enhancement was being made evolving 40-pin DIP to 68-pin grid array (PGA), 387 staggered PGA (SPGA). Today modern CPUs are primarily of two types: Socket type and Single Edge Contact (SEC). Intel CPU's prior to the Pentium II were Socket 7-8, as are the CYRIX and AMD CPUs. The Pentium II and some P-III are SEC and Pentium 4 is Socket 478 CPUs.

The release of Intel 4004 was a good start to produce Intel processor model 8080 which gave birth to what we usually call the **first generation computer chips, intel-8088** and **8086**. The Intel 8088 chip was an 8-bit microprocessor released in 1979, and it was the first chip used in IBM personal computers, though the Intel **8086** chip which was a 16-bit microprocessor, was released earlier, in 1978. The reason was purely economic, since 8088 operated well on cheaper motherboards. Its 8-bit data bus required less costly motherboards than the 16-bit 8086. The first generation 8086 was more powerful and its production became innovative turning to the x86 series, which we now use in modern computers

Below is the list of names of the CPU models as they progressed in processing power and speed over time:

- o 8088
- o 8086
- o 80286 popularly known as the *286*
- o 80386 popularly known as the *386*
- o 80486 popularly known as the *486*
- o 80586 popularly known as the *Pentium (5x86)*
- o Pentium MMX
- o Pentium Pro
- o Pentium II
- o Pentium III
- o Pentium IV

The reader must bear in mind that this list is not exhaustive. Different versions of each model have come and gone, and a few examples of such models are the 386-SX and 486-DX. Usually Intel will modify the model name if the speed of the CPU increases, but the internal makeup of the chip does not change much. For the 486 DX was a 486 chip that was twice as fast as the original 486. Another example is the Intel Pentium MMX CPU. Seeking to resolve compatibility issues among the various add-on multimedia hardware and multimedia enhancement, the Intel company placed the multimedia components directly inside the CPU and called it MMX. Though Intel Corporation owns most of the personal computers CPU market, there is keen competition from companies such as Cyrix, AMD, and others that causes the PC prices to go down continually. Nowadays, in the Pentium era, there are other CPU models such Intel *Celeron*, AMD *Athlon* and AMD *Duron* on the market. You should also be aware that there are several brands and "generations" of CPU chips. The most common are probably the Intel Pentium, AMD-K series, CYRIX-Mx series, and Motorola microprocessors used in Macintosh.

Modern CPUs are **downwardly** compatible with earlier CPUs of the same series, which means that they can run programs designed for earlier CPUs. For example, Pentium Pro can run programs written for all microprocessors from 80x86 series.

The table **3.3** below shows the generations of Intel microprocessors from first generation 8088/86 in the late 1970s to the seventh-generation Pentium-4, released in late 2000. The table has seven columns, including, processor type/generation, year it was released, data bus width, addressing bus width, level 1 cache, memory bus speed, and Internal clock speed.

Table 3.3. Generations of Intel microprocessors from 8088/86 to Pentium- 4

Gene-ration	Type	Year	Package Type	Word Size	Data/ Address bus width	Level 1 Cache (KB)	Memory bus speed fsb (MHz)	Internal clock speed (MHz)
Incub-ation	4004	1971	16-pin Dip	4-bit	4-bit	None		740KHz
	8080	1974	40-pin Dip	8-bit	8/16 bit	None		1.3-2
First	8088	1979	40-pin Dip	16-bit	8/20 bit	None	4.77-8	4.77-8
	8086	1978		16-bit	16/20 bit	None	4.77-8	4.77-8
Second	80286	1982	68-PGA	16-bit	16/24 bit	None	6-20	6-20
Third	80386DX	1985	132-PGA	32-bit	32/32 bit	None	16-33	16-33
	80386SX	1988	132-PGA	32-bit	16/32 bit	8	16-33	16-33
Fourth	80486DX	1989	168-PGA	32-bit	32/32 bit	8	25-50	25-50
	80486SX	1989		32-bit	32/32 bit	8	25-50	25-50
	80486DX2	1992		32-bit	32/32 bit	8	25-40	50-80
	80486DX4	1994		32-bit	32/32 bit	8+8	25-40	75-120
Fifth	Pentium	1993	273-296	64-bit	64/32 bit	8+8	60-66	60-200
	MMX	1997		64-bit	64/32 bit	16+16	66	166-233
Sixth	Pentium Pro	1995	387-SPGA	64-bit	64/36 bit	8+8	66	150-200
	Pentium II	1997		64-bit	64/36 bit	16+16	66	233-300
	Pentium II	1998	242-SEC	64-bit	64/36 bit	16+16	66/100	300-450
	Pentium III	1999	Sock-370	64-bit	64/36 bit	16+16	100	450 - 1.2GHz
Seventh	AMD Athlon	1999	462-pin	64-bit	64/36 bit	64+64	266	500 - 1.67GHz
	Pentium 4	2000	Socket 478	64-bit	64/36 bit	12+8	400/533	1.4GHz -2.6GHz
Eighth	Pentium 4	2002	Socket 775	64-bit	64/128-bit	2MB	800/604	3.2GHz 4.0GHz

The last five columns describing data/address bus, cache level, bus and clock speeds will be discussed further down this chapter. Needless to panic for now, since all information in the table will definitely make sense as you complete this chapter. The first two rows of the table are designated as the incubation period. During this period the Intel 4004 and 8080 microprocessors, which represent today's original **CISC,** architecture were released.

DECIDING WHICH PROCESSORS IS SUITABLE

With the processor features so far discussed above, we now would be able to decide which processor brand would be our best choice. It's a simple choice between two of the industry's giants. The current processor innovative force of the IBM compatible personal computers is undisputedly now driven by two PC processor manufacturing giants. Each company creates new breed of processors, almost every year, which usually compete with each other in terms of processor speed on today's PC market. Currently, Intel markets four ranges of processors while AMD can boast of three popular chips. The *table 3.4* below shows the equivalence of Pentium IV chips in AMD brand. The first column lists four popular Intel CPUs, while their competing breeds produced by AMD are listed in the second column

Table 3.4. Shows the equivalence of AMD Processors to Intel's

Current Intel Processor	Current AMD Processor
Celeron	Athlon XP - Socket A
Pentium IV (478-pin)	Athlon 64 - Socket 754
Pentium IV LGA 775 (775-land)	Athlon 64FX "Opteron" - Socket 940
Xeon (Server)	

The **Celeron** and **Pentium IV** processors use the same Socket 478 motherboard connection and are generally cross-compatible in regard to the motherboard it's being used on. **Xeon** processors use their own socket of either 603 or 604 pins. T

There is another new breed of Pentium IV now available, to support the new technologies of **PCI-Express** (*a faster transport for add-on cards such as graphics cards and high-speed networking*) and **DDR-2** (Double Data Rate RAM Version 2). The new Pentium chip comes with 775 flat connectors which **"land"** on a grid of motherboard pins, hence known as "*land-grid-array, 775 connectors*" or "LGA775" and so will only fit onto the new 775-land motherboards.

Celeron: Celeron processors are usually stripped down variant of their equivalent standard Pentium processors. They are aimed at the low budget PC market, as they are significantly cheaper than their standard Pentium equivalents. The range of Celeron processor clock speeds of 1.8GHz to 2.8GHz, is generally slower than the Pentium IV range, there is some overlap. The significant difference is the size of the **L2 cache**. For all Celerons, this is set at **128KB.** At a cheap price, the Celerons provide more than adequate performance for the average home or office use.

Pentium IV

Figure 3.10 Intel P4 Sticker

In most brand new computers carrying Intel chips you will most likely find Intel sticker like the one in figure 3.11 placed on it

Pentium IV processors are the mainstream processor series marketed by Intel. With clock speeds between 2.4GHz and 3.6GHz, they generally run faster than the Celeron breed, but more importantly have an L2 cache of between 512KB and 1MB. As discussed earlier, a larger cache size makes a tremendous difference to processor performance under heavy application loading, such as games, video editing or engineering software.

Xeon processors are aimed at the server market and come in a number of different versions, with bus speeds of 400/533/800MHz and cache sizes between 512K and 2MB. The older Xeons running at 400MHz use a 603-pin socket, whilst the newer variants running at 533MHz and 800MHz bus speeds use Socket 604. Due to the different architecture in Xeon chips compared with the Pentium CPUs, Xeons produce significantly better performance for a given clock speed than their desktop CPU equivalent, and are designed to operate particularly well when doubled-up, or even quadrupled on a single board, such that very high server loading can occur, with the individual processes being doled out to the different processors.

AMD PROCESSORS - ATHLON

Figure 3.12. The Logo of AMD 64 Athlon.

AMD Processor is another breed of CPU equivalent to Intel Pentium processor. Some computer experts believe AMD processors run faster than Intel Pentium equivalents.

AMD Processors have evolved using very different architecture to those of Intel, there is no direct equality of performance between the two manufacturer's chips for the same given clock speed. In actual fact, an Intel processor operating at **2.8GHz** will not process a given set of instructions as quickly as an AMD processor running at the same speed. The **AMD chip** equivalent in performance and price of manufacture to a **2.8GHz Pentium IV** would run at only **2.0GHz**. In other words a **2.0GHz AMD chip would run the same speed as a 2.8GHz Pentium IV** processor. This does not necessarily mean that Intel processors are not as good as **AMD,** but they are just different.

When buying a computer now the clock speed would normally be the prime headline in making purchasing decisions. **AMD** followed the steps of the Intel chips manufacturing trend, and started producing slower clock speeds, although giving parity of performance over a higher clock-speed Intel chip, would be seen as an inferior marketing tool. Knowing no other information, for example, which processor would you rather buy a **2.0GHz AMD**, or a **2.8GHz Intel chip**? **AMD** responded to this by matching their processor performance indices to that of Intel processors. Given Intel MHz equivalents for example, a **1.8GHz Athlon XP** is marketed as a "**2200XP**", since it would be equivalent in performance to an Intel chip running at **2.2GHz** (**2200MHz**). The indices are used on both the **Athlon XP** and **Athlon 64** range. **AMD** currently market three types of processor, each with their own motherboard socket types:

1. Athlon XP - Socket A
2. Athlon 64 - Socket 754
3. Athlon 64FX ("Opteron") - Socket 940

Athlon XP processors have been the mainstream **AMD** processor line for some years now. Whilst they are expected to be replaced soon in favor of the **Athlon 64**, support will continue for quite some time, and they have the advantage of mature and proven technology, combined with excellent value. They still make a good choice for the budget conscious buyer, and perhaps represent the best bang-for-your-buck of any processor currently available. They are currently available in indices of **2200** to **3200** (**2.2GHz** to **3.2GHz Intel P4** equivalent). Most come with a **512KB** L2 cache. AMD do not plan to develop the XP beyond the **3200** mark.

The **Athlon 64** processor, released earlier this year is AMD's next generation CPU. It is the first mainstream processor to work with both **32-bit** and **64-bit** wide instructions, although most for the vast majority of users, this capability will not be needed, as only a handful of applications at this time have been written for **64-bits**, and only one, specialist, version of Microsoft windows currently supports 64-bit instructions. This chip, however, will work normally, and very well, using **32-bit** software.

More significantly, perhaps, the Athlon 64 it is also the first desktop processor to incorporate the memory controller hub (MCH) or "Northbridge" chipset on the chip itself. As discussed in our Motherboards Guide, the Northbridge traditionally resides on the motherboard. By integrating the Northbridge onto the chip the Athlon 64 now has direct access to the system memory, up to a potential equivalent Front-Side-Bus speed (**FSB**) of **2000MHz** Currently, however, AMD have limited this memory bandwidth to **800MHz**. The Athlon64 has also a larger L2 cache than the XP, at **1MB**.

AMD have high hopes for the Athlon 64 and see it as the medium-term replacement for the Athlon XP, for use in desktop PCs, and the direct competitor to the new breed of Pentium 4 Prescott processors in both price and performance. Currently, the chip is available in speeds of 2800 to 3400 (2.8GHz to 3.4GHz Intel P4 equivalent).

The final chip in the AMD line-up is the Athlon 64FX, or Opteron. Whilst much of the architecture is identical to that of the Athlon 64, the Opteron needs a different socket (940), to cope with the Dual-Channel memory interface. Because of their high-price and superior performance under loading, they are aimed at the server market, in competition with the Intel **Xeon** processor.

OVERCLOCKING

Extremely high speed of CPU will make your computer do millions of operations in a very short time. But heat is released when the CPU runs. High temperature, however, may decrease the performance of your CPU. That's why there is a need for a fan that sits inside the computer box to cool down the processor. You can change the speed of the CPU in the BIOS of your computer. However, excessive increase of the CPU speed will not do any good, it will cause *overloading* and *overheating* which will make the whole computer break down.

Take a quick look at the table 3.3 above and scan the word-size and CPU Speed columns. One will wonder why several processors, for example the Pentiums have equal word-sizes but could run at different speed levels. The reason behind this huge difference in clock cycle rates is, perhaps, due to **Over-clocking**. Over-clocking is the practice of running the CPU (or other parts) past the speed that it is rated at. This is somewhat unknown and unpopular to many computer users. An example is running a 1.2 GHz CPU at 1.4 GHz or a 200 MHz CPU at 233 MHz. How can this be achieved? The following description isn't exact, but it captures the basic idea. Most CPU companies create their CPUs and then test them at a certain speed. If the CPU fails at a certain speed, then it is sold as a CPU at the next lower speed. The tests are usually very severe conforming to a set of standard rules, so a CPU may be able to run at the higher speed quite reliably. ***Over-clocking*** can also be achieved by changing the Bus Speed. In this way the computer is safe and more stable.

Some computers come to market already over-clocked. So you must be cautious when shopping for computers with high speed CPU. This is why it is very important to buy from a dealer you can trust. Over-clocking is not limited to the CPU alone. Some expansion cards such as the video cards are also very over-clockable with some computer companies selling their cards already over-clocked (and advertised this way). The Programs like Power-strip can often be used to easily over-clock the cards. Tweaking computer CPU may come with both advantages and disadvantages.

Advantages
The only advantage in over-clocking may lie in speed gain. However, one cannot overlook the potential risk involved in this practice, which may sometimes outweigh the benefits. Therefore your best judgment would be based on whether or not over-clocking may be worth the risk.

Disadvantages
Over-clocking is dangerous. Running your computer at very high speed will generate excessive heat, which can damage your computer. For example over-clocking your computer, and trying to run a 500 MHz CPU at 1 GHz, has a higher risk of damage most of the time. Also, if you start increasing voltage settings to allow your CPU to run at a higher speed, there is more of a risk there.

Solutions

Heat Control is one of the best way, to resolve over-clocking side effects is to avoid overheating to your computer. In this way you will prevent any potential damage to your computer by keeping your CPU as cool as possible. The only way you can really damage your CPU is if it gets too hot. Adequate cooling is one of the keys to successful over-clocking. Using large *heatsinks* with powerful ball-bearing fans will help to achieve this. How hot is too hot? If you can't keep your finger on the CPU's *heatsink* comfortably, then it is probably too hot and you should lower the CPU's speed.

Changing Bus Speed

Changing the bus speed is actually more beneficial than changing the CPU's speed. The bus speed is basically the speed at which the CPU communicates with the rest of the computer. When you increase the bus speed, in many cases you will be over-clocking all the parts in your AGP, PCI slots, and your RAM as well as the CPU. Usually this is by a small margin and won't hurt these components. Pay attention to them though. If they're getting too hot, you may need to add extra cooling for them (an additional fan in your case). Just like your CPU, if they get too hot, they may be damaged as well.

Believe it or not, it's actually quite simple. In many cases all you have to do is change a couple of jumpers on the motherboard or change settings in your motherboard's BIOS.

Choosing Appropriate Speeds.

The over-clocked speed must be proportional to a determinant factor known as a multiplier. It is a constant value that is used to multiply the bus speed to arrive at the desirable CPU speed. Most of today's

CPUs are multiplier locked, but you can change the bus speed. As an example, you could run a **1.2 GHz** Thunderbird that normally runs at **133** bus (also called 266 because it is "double-pumped) at different speeds given by the formula below:

```
9 x 133 = 1,200 MHz = 1.2 GHz = default

9 x 140 = 1,260 MHz = 1.26 GHz

9 x 145 = 1,305 MHz = 1.3 GHz

9 x 150 = 1,350 MHz = 1.35 GHz
```

Even though that CPU is multiplier locked, you can change the multiplier by connecting the "L1" dots on the CPU itself with a normal pencil (it's just enough to conduct electricity to allow you to change the multipliers). If you do this properly, it is perfectly safe. Below are examples of common variables for multipliers.

```
9 x 133 = 1,200 MHz = 1.2 GHz = default

9.5 x 133 = 1,264 MHz = 1.264 GHz

10 x 133 = 1,333 MHz = 1.333 GHz

Or change both together, like this:

10 x 140 = 1,400 MHz = 1.4 GHz
```

As shown above, extra caution is needed at this point. All you need to do here is use a little common sense when raising the value for the multiplier. For example, you wouldn't want to try to run a **233 MHz** CPU at **400 MHz**, because, it's most obvious, that it won't work or, it probably would damage your CPU. It is highly recommended starting out low and go higher slowly. If you have a 233 MHz CPU, try running it one step higher, then the next step. Most likely you won't be able to get a CPU like this to run much higher than 300, but that is a possibility.

It makes sense to be more concerned with changing the bus speed than the CPU speed, as the former will provide the greatest amount of speed improvement. For example, running a CPU at 250 (83.3x3) would be better than 262.5 (75x3.5) in most cases because the bus speed of 83 is higher than 75. The default for most CPUs is at 66 MHz bus speed. The newer Pentium II/III bus speed is 100 MHz by default while P4's are **400MHz**. Many computers will not have options on bus speed, so watch out for different bus speed options. The higher bus speed you can run at reliably, the better. Depending on what your other components are though, they may cause your computer to crash or become unstable if they can't handle the higher bus speeds. With bus speeds like 133, you have to have higher quality **PC133** or **PC2100 DDR SDRAM** to be able to achieve this bus speed reliably.

Exercise 3.

Select the most appropriate answer to each one of the questions 1 to 20 from the five alternatives labeled A-E below

1. The term integrated circuit **IC** refers to which computer hardware?

    A. Computer machine.

    B. Computer micro-chip

    C. Computer motherboard

    D. Computer microprocessor and motherboard.

    E. All of the above.

3. What's the technical term given to operations CPU performs?

    A. Data processing.

    B. Execution

    C. System programming

    D. Instruction sets.

    E. Logic operations.

4. The term *CPU* refers to:

    A. The computer itself.

    B. All the physical electronic components.

    C. The intelligent part of computer called processor.

    D. Everything within the system unit box only.

    E. All of the above

5. How does apple Macintosh differ from IBM computers?

    A. They come from different manufacturers

    B. They have different CPU architectures

    C. They process data at different speeds

    D. Macintosh are more powerful than IBM computers

    E. IBM computers can store more information

6. Assuming you are to counsel a friend who wants to buy his/her first computer, which of the following computer components will be most important subject to discuss?

    A. Suitable system unit and the input/output peripherals

    B. System unit and the monitor and the printer.

    C. The CPU and the monitor only.

    D. The microprocessor, motherboard, monitor and printer.

    E. The microprocessor, CPU, System Unit and monitor.

7. When one program can run on two different computer brands, these computers are said to be?

    A. Personal computers

    B. Compatible

    C. Over-clocked

    D. Emulators

    E. IBM computers

8. In the basic architecture of CPU of a standard PC, which of the following components is responsible for arithmetic operations?

   A. control

   B. datapath

   C. memory

   D. output

   E. input

9. What is the major determinant factor for CPU compatibility

   A. Instruction sets

   B. Program execution

   C. CPU clock cycle rates

   D. The CPU word-size for basic programs

   E. Data processing bus.

10. The first version of Disk Operation System (DOS) created by Microsoft came to run on Intel chip:

   A. 8088

   B. 8086

   C. Apple II 6502

   D. 80286

   E. 80386

11. For a typical PC system unit architecture the basic components commonly include the following.
    A. CPU, Memory and I/O interface.
    B. Microprocessor and, I/O interface.
    C. Control/Datapath, and Memory.
    D. The integrated circuit and memory
    E. The CPU, Motherboard, and memory.

12. Which of the following is a **major** factor to distinguish between two different CPU models of the same brand?
    A. speed
    B. storage capacity
    C. price
    D. intelligence
    E. Compatibility

13. The internal communications between the CPU and its peripherals occur via which device?
    A. Modems
    B. Ports
    C. Bays
    D. Slots
    E. Bus

14. Which of the following holds the program execution unit of CPU

    A. Core

    B. Integer ALU and Registers

    C. Code Cache

    D. Data Cache

    E. Instruction decode unit

15. To connect your PC's system unit to its peripherals for inputs and outputs, and other communications, your PC must have?

    A. ports

    B. bays

    C. slots

    D. cards and adapters

    E. modems

16. What's Accumulator in a CPU?

    A. It is CPU data bus

    B. It is a permanent data buffer

    C. It is CPU register

    D. It is CPU Cache

    E. It is Branch Predictor

17. To process data CPU understands only **two letters** which are.
    A. 0=*OFF* and 1=*ON*
    B. 1=*OFF* and 0=*ON*
    C. 1=*OFF* and 2=*ON*
    D. 10=*OFF* and 01=*ON*
    E. 00=*OFF* and 11=*ON*

18. The signals generated from your PC's standard keyboard for the letter "**A**" will be equal to:
    A. *OFFON OFF OFF OFF OFF ON*
    B. *OFF OFF ON OFF OFF ON OFF*
    C. *ON OFF OFF OFF OFF OFF ON*
    D. *OFF OFF OFF OFF ON OFF ON*
    E. *ON OFF OFF OFF OFF ON ON*

19. Which of the following best represents the two most popular standards for PCs character coding?
    A. ASCII 8-bits and EBCDIC 8-bits
    B. ASCII 8-bits and EBCDIC 7-bits
    C. ASCII 7-bits and EBCDIC 16-bits
    D. ASCII 7-bits and EBCDIC 7-bits
    E. ASCII 7-bits and EBCDIC 8-bits

20. Using your response in question 16 and 17 above derive the binary equivalence (*bits*) for the letter "A" typed in ASCII standard

    A. 00100001_2

    B. 11000001_2

    C. 10000001_2

    D. 01100001_2

    E. 01000001_2

21. Computer A, which has a CPU speed of 300 MHz takes about 5 seconds to read some instructions. We want to purchase a computer B that will read the same instructions in 3 seconds under equal environmental conditions. We also want Computer B to be at least 1.2 times faster than Computer A. Calculate the speed for computer B?

    A. 600 MHz

    B. 500 MHz

    C. 460 MHz

    D. 500 MHz

    E. 360 MHz

Computer Memory

Chapter 4

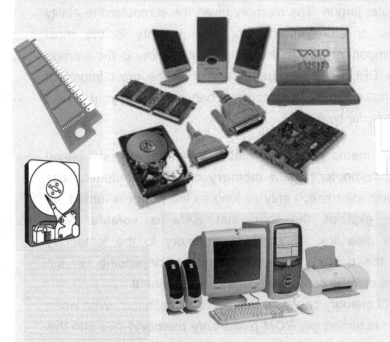

- o **The Computer Memory**
- o **Memory Types**
- o **Connectors For RAM**
- o **Memory Speed**
- o **Memory Capacity**
- o **Cache and Virtual Memory**
- o **Upgrading your Computer**

The Computer Memory

Generally, the word memory refers to the ability to *remember* or *retain* information. The same meaning of memory is somehow used in the computer jargon. The memory gives the computer the ability to remember or store information. The memory is the most important component next to the CPU. The memory is the *storage tank* for the CPU. The storage capability is the most important feature that makes your PC computer a better tool than your pocket calculator and your typewriter.

Some types of memory store information for a relatively brief period of time. The computer's **main memory** or RAM (random-access memory) stores information only as long as the power is turned on. Furthermore, explains the term that RAM is **volatile**. Most programmers also link the computer memory to the temporary memory or the primary memory. Therefore, depending on the operation, the storage can be *temporary* or *permanent*.

Other types of memory can store information indefinitely, even when the computer is turned off. **ROM (read-only memory)** falls into this category so do hard drives, floppy disks, and other storage media. Actually, in the PC world we can easily identify two categories of memory, temporary that is *primary* or main and the permanent also known as the secondary. The main memory is where programs are stored along with its data as it executes. The temporary memory and the ROM are mostly *integrated circuits (/C) or chips* whereas the permanent memories are *physical devices*.

Memory Types

ROM

As mentioned earlier ROM is read-only memory. ROM is read-only memory. Unlike the RAM, ROM is non-volatile memory and is therefore used to hold data permanently. Data is very secure since it is "hard-wired" into the ROM chip so you can store the chip forever and the data will always be there. The BIOS is stored on ROM because the user cannot disrupt the information. ROM is slower than RAM, which is why some try to shadow it to increase speed. **ROMs** usually are memory chips that have their stored content entered at the time of fabrication. It can be written to only once, and content is not lost if power is disrupted.

Types of ROM

In the computer market now, we can count three different types of ROM namely **PROM** *(Programmable ROM)*, **EPROM** *(Erasable PROM), and **EEPROM** *(Electrically EPROM)*.

Programmable ROM (PROM) This is basically a blank ROM chip that can be written to once. It is much like a CD-R drive that burns the data into the CD. Some companies use special machinery to write PROMs for special purposes.

Erasable Programmable ROM (EPROM) EPROM is similar to PROM except the ROM can be erased by shining a special ultra-violet light into a sensor on the ROM chip for a certain amount of time. This process will erase data on the ROM chip allowing the chip to be rewritten.

Electrically Erasable Programmable ROM (EEPROM) (Also called Flash BIOS) This ROM can be rewritten through the use of a special software program. Flash BIOS also operates this way, allowing users to upgrade their BIOS. Most modern PC's come with EEPROM making its BIOS upgradeable.

Primary and Secondary Memories

In modern computer architecture the RAM is referred to as the *primary memory*, while the permanent storage device such as the hard-drive refers to the *secondary memory*. These names were derived from the process through which computer systems look for data internally. The process of looking for information starts from the *Cache* followed by the RAM in case of the absence of data in the cache. Failure to find information in the primary memory is termed *page fault*. In addition, primary memory is the location of the first search attempt or *primary*. In the event of a page fault the computer system then refers to the permanent storage device, which may be a hard-drive or floppy drive. This being a second attempt and as the target is the storage device it sounds therefore so natural to name it *secondary memory*. If at this point the data cannot be found, the computer system may thence confirm the absence of that particular data in a form an error message.

Primary Memory or (RAM, CACHE)

The temporary storage device is what most programmers call memory, and which is also known as random-access memory or RAM. The RAM is the temporary data holding tank inside the computer system where the CPU draws whatever is being processed. The memory is the location of stored programs as well as data necessary to run the programs. The amount of RAM of a computer system determines the number of programs that can run simultaneously without disrupting the system performance, and how well each application will perform. Usually the RAM is pre-installed in most computers. Your computer's operating system requires a good deal of RAM to run smoothly. Generally, all things being equal, more RAM implies an more powerful CPU. Modern computers are equipped with another temporary storage memory known as *Cache* serving the same purpose but it's used as the buffer for the RAM. Cache memory holding data temporarily may lead to faster access time. Like the RAM, more cache means a more powerful CPU.

Like the CPU, RAM comes in different flavors, which pose problems to newcomers when looking to buy RAM on the market. You have to be sure to buy the right type for your particular computer. The chapter will guide you to the critical specifications.

Types Of RAM

Competition in the PC world is such that each computer component evolves over time to match the growing pace of the CPU evolution. The RAM is no exception from this continual evolution. New computer systems should come with matched versions of RAM. Overwhelming breed of different memory chips often make identification of RAM types look somehow cumbersome. When you visit the Computer store instead of hearing RAM you will probably hear terms like **DRAM, SDRAM,** *SIMMs, DIMMs and RIMMs*. Now to be able to classify a RAM you need to know four things:

- *capacity*,
- *type*
- *connector*
- *speed*.

The capacity of RAM will be discussed further in this chapter, but for now, let's discuss the types of RAM. You may have seen a variety of RAM types on the market, but technically there are only two basic types that can be identified as *static* or *dynamic.*

Static RAM (SRAM)

This RAM will maintain its data as long as power is provided to the memory chips. It does not need to be rewritten periodically. SRAM is very fast but much more expensive than DRAM. SRAM is often used as cache memory due to its speed.

Dynamic RAM (DRAM)

DRAM, unlike SRAM, must be continually rewritten in order for it to maintain its data. This is done by placing the memory on refresh circuit that rewrites the data several hundred times per second. DRAM is used for most system memory because it is cheap and small. Modern technology has a big challenge of boosting the speed of computer system. You have had a greater share of this booster when discussing the CPU. Much attention has been focused on improvement of Dynamic RAM's.

Types of Dynamic RAM (DRAM)

There are several types of DRAM flooding the PC market and complicating the memory scene even more. The latest and fastest

RAM on the market now is **RD** or **DDR** (*Rambus Dynamic and Double Data Rate*). You will also find earlier versions like the **EDO** (*Extended Data Out*) and SD (*Synchronous Dynamic*) RAM. The SD RAM is also called "**SDRAM**" (*Synchronous Dynamic Random Access Memory*). Hardware engineers have proven SDRAM faster in access than DRAM. SDRAM was designed for more memory space and speed. Also, since images usually consumes more memory space there has been a need for graphics memory like SGRAM or *Synchronous Graphics Random Access Memory*. Now let's discuss each of the Dynamic RAM.

EDRAM - Enhanced Dynamic Random Access Memory is a form of DRAM that boosts performance by placing a small complement of static RAM (SRAM) in each DRAM chip and using the SRAM as a cache. Also known as *cached* DRAM, or **CDRAM.**

Fast Page Mode DRAM (FPM DRAM)

FPM DRAM is only slightly faster than regular DRAM. It uses a slightly more efficient method of calling data from the memory. FPM DRAM is not used much anymore due to its slow speed, but it is almost universally supported.

Extended Data Out DRAM (EDO DRAM)

EDO memory incorporates yet another tweak in the method of access. It is a form of DRAM that has a two-stage pipeline, which lets the memory controller read data off the chip while it is being reset for the next operation. It allows one access to begin while another is being completed. While this might sound ingenious, the performance increase over FPM DRAM is only slight. EDO DRAM must be properly supported by the chipset, but it is the most common type of memory for most users. Power users with high bus speeds typically opt for something faster, though. While similar in performance to synchronous DRAM (SDRAM), it cannot support bus speeds above 66MHz.

Burst EDO DRAM (BEDO DRAM)

This is basically EDO DRAM with combined pipelining technology. The result is a much faster EDO memory chip capable of working with faster bus speeds. Support for the BEDO technology is rather sparse. SDRAM has caught on faster. *Bursting* is a rapid data-transfer technique that automatically generates a block of data (a series of consecutive addresses) every time the processor requests a single address. The assumption is that the next data-address the processor will request will be sequential to the previous one. Bursting can be applied both to read and write operations to and from memory.

Synchronous DRAM (SDRAM)

SDRAM is the developing new standard for PC memory. Its speed is synchronous, meaning that it is directly dependent on the clock speed of the entire system. It works at the same speed as the system bus, up to 100MHz. Although the SDRAM is faster BEDO DRAM, the speed difference is not notice by many users due to the fact that the system cache masks it. Also, most users are working on a relatively slow 66MHz bus speed, which doesn't use the SDRAM to its full capacity.

RAMBus DRAM (RDRAM)

Usually known as Direct Rambus DRAM (**DRDRAM**). This is a technology still being developed by Intel that may surpass SDRAM. Its goal is to get rid of the latency, the time taken to access memory. It does this by actually narrowing the bus path and treating the memory bus as a separate communication channel. The Direct Rambus DRAM is a totally new RAM architecture, complete with bus mastering (the Rambus Channel Master) and a new pathway (the Rambus Channel) between memory devices (the Rambus Channel Slaves). A single Rambus Channel has the potential to reach **500 MBps** in burst mode; about 20 times increase over DRAM.

Memory Parity

There is another important feature that you must watch out for when shopping for RAM. The type of Parity on RAM can be a concern. Parity is simply a kind of error correction method during data transmission. Though most PCs use non-parity memory, older systems may use parity memory. Business servers and high-end PCs use ECC (*error correction code*) memory. Parity memory works in non-parity systems but not the vice versa. ECC memory is capable of detecting and correcting 1-bit errors. Depending on the type of memory controller ECC can also detect rare 2-, 3-, or 4-bit memory errors. However, only single-bit errors can be corrected. In the case of a multiple-bit error, the ECC circuit reports a parity error. ECC logic may be included on a DIMM/SIMM, or it may be found on one of the computer's circuit boards. On a SIMM or DIMM, the use of a module addressing architecture that facilitates the use of the memory module by systems with ECC is termed *ECC optimized memory*. ECC optimized memory modules do not have byte-write capability.

Connectors for RAM

Below is an illustration of three types of memory modules that carry the memory chips described above. Remember that RAMs are memory chips and memory chips are also integrated circuits.

Figure 4.1

RAM (Random Access Memory) or SIMM carrying 8 chips

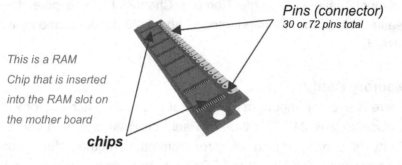

Pins (connector)
30 or 72 pins total

This is a RAM
Chip that is inserted
into the RAM slot on
the mother board

chips

Figure 4.2

168 *Pins (conne*

RAM (Random Access Memory) on DIMM carrying 8 chips

Memor
chips

Figure 4.3   Shows RIMM Memory carrying 184 pins

Memory chips

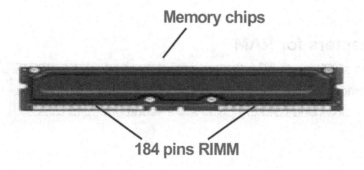

184 pins RIMM

RAM comes in *modules* loaded with *storage cells*. These storage cells or *memory chips* reside on modules that have connecting pins which slots into memory sockets on the motherboard. There are two main types of memory modules: SIMM (*Single-In-Line Memory Module*), and DIMM (*Dual In-Line Memory Module*). Under dual in-line memory module, there is a new version developed by a company called Rambus. Rambus version is known as RIMMS (*Rambus In-Line Memory Module*). The SIMM shown in figure 4.1 has *72 connecting pins*, DIMM has *168 pins*, and RIMM has *184 pins* as shown in figures 4.2 and 4.3 respectively.

Unlike the dual modules, SIMMs must be grouped together in pairs, or "*banks*" where each module has the same capacity. The mother-board of a typical older Pentium PC would have four sockets, equal to two banks, so possible configuration would include:

Table 4.1 Possible configuration for SIMMs

	OPTION 1	OPTION 2	OPTION 3
1st bank	2 x 8MB SIMMs	2 x 16MB SIMMs	2 x 16MB SIMMs
2nd bank	2 x 8MB SIMMs	2 x 16MB SIMMs	2 x 32MB SIMMs
Total RAM	32 MB	64 MB	96 MB

DIMMs have no restrictions, since the module is already dual or paired by the manufacturer. In certain computer systems higher capacity memory modules like DIMM may be pre-installed. DIMMs support 64-bit and higher buses and have 168 pins, and RIMMs are superfast designed with 184 connecting pins. Notebook computers (laptops) usually use SoDIMMs (Small-outline DIMMs), which are normal DIMM smaller in size. Like the big DIMMs, one SoDIMM makes a complete bank.

RAM SPEED

The RAM modules communicate with the processor at different rates (commonly 66, 100, and 133MHz.) This is a important factor since it directly affects the speed of the chipsets on the motherboard. Matching suitable speed of the RAM to that of the CPU will be beneficial.

Synchronous Graphics Random Access Memory SGRAM and (AGP, VRAM)

The *SGRAM* has the same structure as what has been described above, but can be only used as video memory. This memory interacts with the software and affects the display because it assists the video card in communicating to the processor (*note CPU*) and the monitor. When it comes to working with graphics and video applications SGRAM becomes inevitable. Today's systems average 4MB to 64MB and above of video memory. Especially in two-dimensional (2-D) and 3-D applications, the SGRAM is needed to perform the maximum resolutions (e.g. 2048 x 1536 W/TVOUT).

Modern computers have an Accelerated Graphics Port or AGP that enables the graphics controller to communicate to the computer and access its main memory. The AGP is a high-speed port designed to handle 3-D technology. In addition, it stores 3-D textures in main memory rather than the video memory. A video adapter may have its own processor known as *Graphics a Accelerator* to enhance performance. These accelerators are designed to handle high-end graphic activities, leaving the CPU to handle other system requirements. Accelerators typically use conventional DRAM or a special type of video RAM called *VRAM* to accommodate the video circuitry and the processor with accessing the memory simultaneously. Some computers have pre-installed video RAM or VRAM and the *Cache* memory already mentioned in the previous topic.

Memory Capacity

RAM is an important device when it comes to running programs on the computer. Proper running of programs may depend on the amount of RAM available to keep all the running information. Programs usually run faster when there is sufficient RAM capacity. Each unit of information is stocked in the memory of the computer in a form of a set of *bits* which stands for *BI*nary digi*Ts*, i.e. it takes values **0** or **1** exclusively. Below is one byte representation with number of bits counting from right to left.

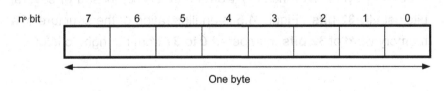

n° bit	7	6	5	4	3	2	1	0

One byte

Figure 4.4.   Bits representation of one byte

RAM is measured in bytes. One (1) byte is equivalent to a set of 8 bits. Generally, 1 byte corresponds to the size of memory space that can hold 1 character. More powerful computers operate well on RAMs capacities measured in megabytes (MB) and gigabytes(GB). The capacity of a computer memory is measured in powers-of-two increments, **4, 8, 16, 32, 64, 128, 256** and **512**MB. Keep in mind that there is always a little extra space added to the RAM when measuring capacity. Actual size is measured in multiples of 2^{10} bytes being *1024* bytes or *1 kilobytes* (*1Kb*). **K** is **1000** and an additional **24** bytes. In micro-computers memory capacity may range from 64Kb to several megabytes (1 Mb = $2^{10} \times 2^{10}$ bytes = 1024Kb). Nowadays, with the rapidly growing pace of technology, it is possible to find RAM chips of capacities rising above 512 MB.

Considering a personal computer with a memory capacity of *256 kilobytes*, that is *1024 * 256 = 262144* characters, and instructions capable of being stored in the memory. To keep records of only character data type of 80 characters long, we can place 3200 records. Also, the manipulation of numeric data type only will permit the storing of over 65000 numbers, knowing each number occupies 4 bytes in memory.

Computers with high memory capacities are measured in *words* instead of bytes. The memory words may be composed of several bits usually **32 bits.** Figure 4.5 is an illustration of the structure of a memory word of **32 bits**, numbered *0* to *31* from the right to left.\

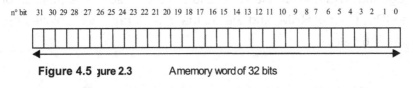

n° bit   31 30 29 28 27 26 25 24 23 22 21 20 19 18 17 16 15 14 13 12 11 10 9 8 7 6 5 4 3 2 1 0

Figure 4.5 ꝑure 2.3   A memory word of 32 bits

Figure 2.3 illustrates structure of memory word composed of 32 bits

Today's computer systems may carry an amount of RAM ranging from 4MB (4 Megabyte or 4 million of bytes) through 32MB to even 2.4 GB (where1 Gigabytes 1 billion bytes). At present time, due to the fast growing pace of technology 16MB RAM may cost about $28 or less.

A PC's operating system will require a good deal of memory to run smoothly. For example, Windows 95 needs at least 8MB to work at all, 16MB to work properly and double the memory to work smoothly. Windows 98 demands at least 16 MB to wake it up and 64MB to run at realistic minimum speed. Windows ME runs only slowly in 32MB. Windows ME does better in 64MB, but really require 128MB to run properly. Windows 2000 works well with 64MB but is also much better with 128MB. Finally, Windows XP operating system might require 256MB to run smoothly.

Cache and Virtual Memory

Computer experts searching for solutions to speed up the computer processing time discovered what we now term as **Cache memory**. Most cache memories of modern computers are built-in to the central processing unit, in a specific location making it possible to bridge the link between the RAM and the processor. The position of the CPU, Cache, RAM, and the permanent storage will look like the figure 4.6 below:

Figure 4.6. The distant Relations of Memory Devices To CPU

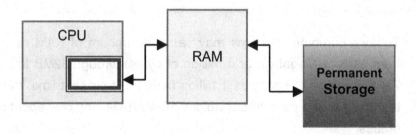

The Cache memory consists of a small fast memory that acts as a buffer for the CPU. It is a small high-speed storage area, which houses frequently accessed data. This means that cache memory is used to hold data temporarily, providing faster access time for the CPU. It is usually attached to the CPU and may have different levels. For example, Level 1 Cache, or L1 Cache, L2 cache etc. Cache actually stores the addresses of the data stored temporarily in the system's memory, so during data research in the system the CPU will first check the cache rather than searching through the entire system. The search sequence follows from the cache through the RAM to the permanent memory until the data is found or confirmed not available. Cache enhances performance by decreasing memory access time. A Pentium III, 866 MHz without cache will not run as quickly as you would like . Today, cache in

Pentium CPU may range from 128 KB, 256 KB, 512 KB to 2 MB. All Pentiums I, II, III and IV CPUs have at least 512 KB cache on a chip. The Xeon P3-9000 processors are said to carry 1 MB to 2 MB L2 cache on them. Here again, more cache would imply more power to your computer.

Virtual Memory

Keep in mind to be able to relocate data, the memory storage areas is divided into storage cells labeled with contiguous addresses. The bigger the memory size, the higher the number of memory addresses made available for storage. At times, your system might run out of memory (i.e. run out of address), which induces RAM overload. At this point the PC has four choices:

- give up the ghost and crash
- freeze
- refuse to do another thing until you reboot
- close down some programs; or add more memory

New versions of operating systems like Microsoft Windows have a better way to resolve this problem without adding physical memory to your system. This method is the virtual memory technique.

The **virtual memory** is an imaginary memory area supported by some operating system (for example, Windows but not DOS) in conjunction with the hardware. It's not a physical memory chip. It's more software than hardware. You can think of virtual memory as an alternate set of memory addresses. Programs use these virtual addresses rather than real addresses to store instructions and data. When the program is actually executed, the virtual addresses are converted into real memory addresses.

As a computer user, it is recommended to maximize the resources available. The process of preparing plans to coordinate the total system to operate in the most efficient manner is called a system *optimization*. **Virtual memory** computing is one of the best methods for system optimization.

The purpose of virtual memory is to enlarge the address space, which is the set of addresses a program can utilize. For example, virtual memory might contain twice as many addresses as main memory. A program using all of virtual memory would not be able to fit in main memory all at once. Nevertheless, the computer could execute such a program by copying into main memory those portions of the program needed at any given point during execution. For example, 32-bit computer will be able to create a maximum size of 2^{32} bytes of virtual addresses:

$$2^{32} \text{ bytes} = 4,294,967,296 \text{ bytes} = 4 \text{ Gigabytes.}$$

At the time of the writing of this book, no personal computer has a memory capacity of 4Gigabytes. The operating system divides virtual memory into **pages**, each of which contains a fixed number of addresses to facilitate copying virtual memory addresses, which are converted into real memory addresses. Each page is stored on a hard disk until it is needed. When the page is needed, the operating system copies it from the disk to main memory, translating the virtual addresses into real addresses.

The process of translating virtual addresses into real addresses is called *mapping*. The copying of virtual pages from disk to main memory is known as *paging* or **swapping**.

Some physical memory is used to keep a list of references to the most recently accessed information on an I/O (input/output) device, such as the hard disk. The optimization it provides, is that it is faster to read the information from physical memory, than using the relevant I/O channel to access information. As discussed in previous paragraphs, this process is called *caching*. Caching is implemented inside the Operating System.

The only disadvantage of virtual memory is when the situation, called *thrashing* occurs. When the operating system constantly swaps data with data in RAM, more pressure may be exerted on the disk struggling to keep on top of its duties affecting the system performance.

The Caching Principle In Virtual Memory

Caching principle can be used in virtual memory for storing address translations. When address translation, for a virtual page number is used, it will probably be needed again in the near future. Therefore, it would be helpful to have a high-speed table of recently used translations. This special address translation cache, located inside the CPU, is traditionally referred to as a **translation look-aside buffer** or TLB. The TLB only contains page table mappings; that is the virtual memory address and the corresponding physical page address. These high-speed registers are expensive and only a few are included on the chip.

Upgrading Your Computer

There are times when computers cannot run as fast as needed or just cannot handle certain processes effectively due to the shortage of system resources. When the limitations of system resources become a major barrier to achieving your maximum productivity, we often consider the apparent ways of upgrading the system. Ways to upgrade the system include switching to a faster CPU, adding more physical memory (RAM), installing utility programs, and so on.

Now that you know what is Ram, you can try adding more memory to your computer or upgrading the computer. The upgrading of a computer is just the addition of such hardware as the RAM chips or storage devices etc, to existing ones to increase the computer performance. In the case of a memory upgrade, the SIMMs, DIMMs, or RIMMs are also plugged into the connectors of the motherboard. The motherboard connectors are referred to as memory expansion slot. Usually the memory chips loaded as SIMMs, are added in pairs. That is, if you plan to increase your memory by 16 MB RAM, it recommended to use two 8MB Ram chips (2 SIMMs) or four 4MB Ram chips (4 SIMMs), depending on the number of expansion slots available on your motherboard. You must refer to table 4.1 for more details on pairing the SIMMs. The latest development on memory modules would be the Dual in-line memory modules (DIMMs and RIMMs). Using Dimms and RIMMs, one will not worry about pairing. Each memory chip will work just fine.

Exercise 4.

Students are required to select the most appropriate answer from the five alternatives labeled for questions 1- 22 listed below:

1. What is the main purpose of RAM in Computer systems ?
    A. It stores data permanently for the computer.
    B. It is part of the CPU hardware.
    C. It implements the virtual memory in the CPU.
    D. It contains a collection of chips used to process data.
    E. It's storage tank for the CPU during program execution.

2. Which of the following is true about Random-Access- Memory?
    A. Computer's main memory
    B. It is volatile.
    C. It is the permanent storage tank for the CPU.
    D. A and B.
    E. A, B, and C.

3. The computer memory, which stores information indefinitely and can be changed by users, is popularly known as?
    A. Hard drive
    B. RAM
    C. Cache
    D. ROM
    E. All of the above

4. In PCs there are two groups of memory primary and secondary. Which of the following can be identified as secondary?

   A. RAM

   B. Cache

   C. Hard Drive

   D. ROM

   E. Virtual Memory.

5. Which of the following may be usually classified under external peripherals?

   A. Hard drive

   B. Mouse

   C. Floppy drive

   D. CD-ROM drive

   E. Graphic Card

6. The first version of virtual memory in PCs was supported by which Operating System.

   A. MS DOS

   B. MS Windows

   C. Apple OS2

   D. Unix Version 4

   E. IBM DOS

7. When selecting your PC memory which of the following is **<u>NOT</u>** important.

    A. Type

    B. Capacity

    C. Speed

    D. Connector

    E. Number of chips on module

8. When upgrading your PC memory, you must follow the following restrictions.

    A. pair SIMMs only

    B. pair DIMMs only

    C. pair RIMMs only

    D. pair SD RAM only

    E. All of the above

9. Assuming your Pentium PC was loaded with 32 MB RAM filling all the four memory slots, you want to upgrade this computer to at least 64 MB RAM. Which of the following configuration is more appropriate.

    A. 2 x 8MB + 1 x 16MB + 1 x 32MB

    B. 3 x 16MB + 1 x 32MB

    C. 2 x 16MB + 2 x 8MB

    D. 2 x 16MB + 1 x 32MB

    E. 2 x 16MB + 2 x 32MB

10. Which of the memory modules can be used for laptops

    A. SIMMs

    B. DIMMs

    C. RIMMs

    D. SoDIMMs

    E. All of the above

11. The process of translating virtual address to real or physical address is known as

    A. mapping

    B. paging

    C. Swapping

    D. coping

    E. caching

12. The video adapter may have its own processors known as

    A. AGP

    B. Graphic Accelerator

    C. Video Ram

    D. Cache memory

    E. 3-D Technology

13. The memory that resides between the CPU and the main memory is
    A. virtual memory
    B. Cache
    C. SD Ram
    D. DDR Ram
    E. Synchronous Dynamic Ram

14. Choose the role(s) which are vital and implemented by the memory?
    A. Stores data permanently
    B. Stores address of data temporarily
    C. Stores data temporary for running program
    D. It serves as a buffer between CPU and Cache memory
    E. None of the above

15. The cache memory is well known for
    A. Storing the address for data in the Ram
    B. Storing data that matches the data stored in the RAM
    C. Storing the same data stored in virtual memory
    D. Storing part of data store in the hard drive
    E. All of the above.

16. When seeking a particular data which of the following order does the CPU follow?

    i. permanent storage
    ii. Cache
    iii. Ram
    iv. virtual memory

A. i, iv, ii, iii   D. ii, iv, iii, i

B. iii, ii, iv, i   E. iv, ii, iii, i

C. iv, iii, ii, i

17. The main role of the Cache memory is

A. serves as buffer to the virtual memory

B. It increases the access time for data

C. It enhances performance by accelerating access time

D. It enhances the virtual memory access time

E. all of the above

18. Which of the following is the main cause for your computer to freeze.

A. insufficient memory – RAM

B. thrashing

C. insufficient storage cells on the hard-drive

D. all of the above

E. only A and C

19. The virtual memory may be classified under

    A. Hardware   D. Physical Ram

    B. Software   E. ROM

    C. Peripheral

20. The process by which to make maximum use of the system resources available for efficiency is known as:

    A. Upgrading   D. Utilization

    B. Restriction   E. Optimization

    C. Installation

21. The purpose of virtual memory is to:

    A. Upgrade the storage capacity of computer.

    B. Enlarge the address space of the system.

    C. Double the cache memory space CPU.

    D. Double the cache memory.

    E. Manipulate the hard drive.

22. What is the maximum size of virtual address can a 32-bit computer machine running under windows 95/98 generate

    A. 25 kilobytes

    B. 128 megabytes

    C. 4 Gigabytes

    D. 8 Gigabytes

    E. 32 Gigabytes

The Computer Permanent Memory

Chapter 5

o   The Permanent Storage Memory

o   Types Storage Devices

o   Hard Disk Drives

o   Preparing Your Hard Disk Drive for Use

o   External Storage

o   Floppy Disk

o   Compact Disks Read Only Memory (CD-ROM)

o   Backup Tapes

The Permanent Memory

The idea of creating computers was to perform complex calculations and obtain a higher precision in calculation operations. In this way computers operate more or less like a calculator. Moreover, the ability to process and store information makes the computer a unique system and distinguishes it from the ordinary calculator. The permanent memory is a storage device of the computer. Before any permanent memory can store data, the device must be *formatted*. Formatting is a process in which a permanent storage device is prepared internally or sub-divided into manageable cells. Formatting organizes tracks on a diskette resulting in a measurable storage device capable of storing data. Partitioning a disk is prerequisite to formatting. Disk partitioning is the process of defining certain areas of the hard disk for the operating system to use. The areas created as the result of partitioning are known as volume. A volume can be identified as **C:** or **D:** thru **Z:**. Partitioning is fully discussed under *"Preparing Your Disk Drive Ready For Use"*. Some popular hard drive manufacturers include Maxtor, Toshiba, and SeaGate.

Storage Capacity

As abovementioned, you must format your hard disk before you can use it. However, bigger hard disks must be partitioned to logical disks before formatting. Each partition is seen by the operating system as a separate disk. Presently, hard disk capacities are averaging about four to eight gigabytes (**GB**) for a low-end system, and 10 to 60GB for high-end systems.

Types Of Storage Devices

Modern personal computers may be equipped with a variety of permanent storage devices, the Hard Drive, Floppy Disk, External Drives, Cartridge-based Storage, Compact Disc Read Only Memory Drive, or CD-ROM, Digital Versatile Disc or DVD-ROM. The most important role of the storage device is to hold the operating system, other programs and the data which the computer manipulates. Like the RAM, all permanent storage devices can be measured in bytes. Larger volumes can be measured up in kilobytes, megabytes, gigabytes or terabytes. They also have two types of measurable speeds, *rotational* and *interface*.

Hard Disk Drive

The storage device which is set in place inside the computer is known as the hard disk drive. The hard disk drive as shown in figure 5.1 below is another critical part of a PC system. The hard disk drive role is to store all information you enter into the computer.

Figure 5.1 PC Hard disk. This device is fixed internally in the PC systems.

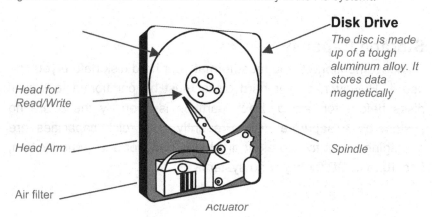

Disk Drive

The disc is made up of a tough aluminum alloy. It stores data magnetically

Head for Read/Write

Head Arm

Spindle

Air filter

Actuator

Figure 5.2 A Labelled Picture of PC Hard disk drive. This device is fixed internally in the PC systems.

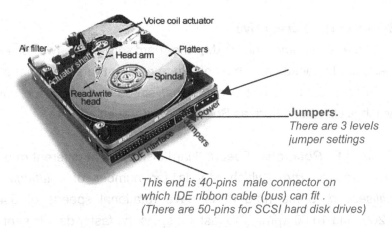

Jumpers.
There are 3 levels jumper settings

*This end is 40-pins male connector on which IDE ribbon cable (bus) can fit .
(There are 50-pins for SCSI hard disk drives)*

Unlike RAM, where data is held in a kind of dynamic flux, files once saved to the hard disk are housed in safe storage. These files can be retrieved from the hard disk, altered or deleted and re-saved.

The hard disks are sealed boxes to store data magnetically on a "hard" inner disk made of a tough aluminium alloy. The hard *drive* is the mechanism that spins the actual disk and moves the magnetic heads to do the hard work. Since the disk and the drive are inseparable in practice, we call the whole assembly a *hard disk*. Hard disk provides permanent storage; that is, it holds your data even when the computer is turned off. Today's computer market may carry a few types of hard disk like **IDE** or **SCSI**(*"scuzzy"*) and **EIDE/ATA/DMA** which will be discussed further below. *IDE* stands for *Integrated Drive Electronics,* and *SCSI* stands for *Small Computer Systems Interface.* **Speed** factor is as important as the capacity when it comes to choosing your hard disk from the list mentioned above.

Speed of Hard disk drive
Like the CPU and the RAM, the hard disk has two different measurable speeds as it has to spin at different rates to either store or retrieve data. It has a *rotational* speed and *interface* speed which is classifced here as Speed #1 and Speed #2.

Speed #1 - Rotational Speed: Hard disk spins at different rates all measured in *rpm,* which stands for number of *rotations per millisecond*. Some hard disks have rotational speeds of 5,400, 7,200, and 10,000rpm. The faster it spins the faster data is sent out to the computer system.

Speed #2 - Interface Speed: Under interface speed, two speeds are noted:
- the transfer rate *in the drives*
- the transfer rates occurring on the cable *connecting* the drive and the CPU.

On the drives, the interface speed is rated in megahertz. Examples are **Ultra 33**, **66**, or **100,** which are sometimes written as **UDMA-33**, UDMA-66 or UDMA-100 respectively. These figures are all in megahertz representing the rate at which the drive could possibly transfer data. The actual data moving to and from the drive and the CPU occur through the ribbon cables with a measuring speed of *megabytes per second* (**MB/Sec**). This is actually the rate at which data is being transferred from the hard disk across the data bus or the 40-pins IDE or 50-pins SCSI ribbon cable to the CPU. Some Intelligent hard drives like **IDE ATA-1** and **ATA-2** can assume data transfer rates ranging between 3.3MB/Sec and 16.6MB/Sec. This part is discussed fully under IDE Interface below.

Preparing Your Disk Drive Ready For Use

Under Storage Capacity, we learned that before using hard drives or floppy drives users must format them. This is a very important aspect of modern computing. There are two types of formatting: **low-level** and **high-level**. Therefore the process of preparing a hard disk for use follows these steps. First, *low-level formats*, then *partitions*, then *high-level formats*. The **low-level formatting** turns the platter or disk from a blank slate to a divided slate. It defines the data areas, creates tracks, separates into sectors, and writes the ID numbers to each sector.

Partitioning is one of the necessary steps to prepare a drive for use. It is the process of defining certain areas of the hard disk for the operating system to use. A *volume* is a section of the drive labeled with a letter, like **C:** or **D:**. Actually, a *volume* may be seen by a computer as a logical area of identification. All hard drives must be partitioned, even if they will have only one partition.

The High Level Formatting sets up the **Disk Operating System** (DOS) structure for the Hard Disk. It writes the boot record, creates the File Allocation Table or FAT table, the directory and tests the Disk for bad sectors. The DOS external command **FORMAT** is used to create the high level format. Disk formatting will be fully discussed under operating systems.

A boot sector is an area on the storage disk which is reserved for the operating system to house system start up data. The computer may not boot up or start without the boot sector. A partition program writes a master partition *boot sector* to cylinder 0, head 0, sector 1.

The data in this sector defines the start and end locations of each of the other partitions. It also indicates which of these partitions is active or bootable, thus telling the computer where to look for the operating system.

How Many Partitions Can Your PC Handle?

All systems can handle 24 partitions, either spread out on the same hard drive or many drives. This means that one can have up to 24 different hard drives, according to DOS (*Disk Operating System*). Some operating systems may recognize more than 24 partitions, DOS will not. The limiting factor is simply the availability of letters. All partitions must have a letter that the operating system can recognize as the *logical name*. There are *26 letters*, but **A:** and **B:** are reserved for floppy drives, leaving 24 letters available. Note it is always *letter + a colon sign*.

Although, there are third party partitioning programs with marketing added capabilities, DOS **FDISK** is the accepted program for partitioning. FDISK sets up the partition in an optimum way and allows more than one operating system (OS) to operate on one system. FDISK only shows two DOS partitions: the *primary partition* and the *extended partition*. The extended partition is divided into logical DOS volumes, each being a separate partition. The minimum partition size is one megabyte due to the fact that FDISK in DOS 4.0 or later creates partitions based on numbers of MB. Partition size is usually limited to 2G. DOS versions earlier than 4.0 allow maximum partitions of 32 MB. The DOS operating system using File Allocation Table (FAT), can handle this partition and promote the allocation of files for storage. Using the FAT32 system under DOS 7 and Windows 95 Operating System Release 2 (<u>Windows 95 OSR2</u>), maximum partition size is increased up to 2 Terabytes (2 TB).

Note that 2 TB = 2048 GB = 2,097,152 MB = 2,147,483,648 KB = 2,199,023,255,552 bytes.

Structure Of Hard Drive

All hard disk drives share the same basic structure, varying only in how each part is used and the quality of the parts themselves. The platters, spindle motor, heads, and head actuator are inside the drive, sealed from the outside. This chamber is often called the head disk assembly (HDA). The HDA is rarely opened, except by professionals. On the outside are the logic board, bezel, and mounting equipment. Unlike the floppy and CD drives, the disks and drives are set in place as one complete assembly. This means you cannot separate the disk from the drive as different devices. The diagrams below illustrate complete head disk assembly.

Figure 5.3. *Hard disk drive exposing the entire head disk assembly*

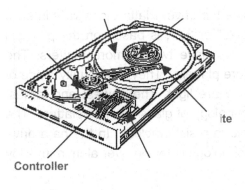

Controller

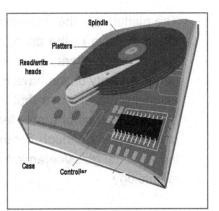

From the above diagram we can see that the major parts inside a Hard Disk assembly include:

- Disk platter or platters.
- Spindle motor for rotating the platters.
- Electromagnetic read/write heads (one per magnetic surface).
- Access arms or armatures from which the heads are suspended
- Actuator for moving the arms (with heads attached).
- Logic board or preamplifier circuitry to maximize read/write signals.
- Air filter and pressure ventilation.

For now we will take a further look at the head disk assembly (HDA) and select the most critical part of the hard disk drive. The most sensitive part is the platters and the read/write heads. The size of surface area of platters will determine the storage capacity as the spinning rate will determine the speed.

The platters are the disks inside the drive. Platters can vary in size. Often the size of the drive, 5.25" or 3.5", is based on the physical size of the platters. Most drives have two or more platters. The larger capacity drives have more platters. They are usually made of an aluminum alloy so that they are light. The newest and largest drives make use of a new technology of glass/ceramic platters. The glass contains enough ceramic to resist cracking. In the hard drive industry, the glass/ceramic technology is taking over aluminum alloy technology.

Figure 5.3. *A hard disk drive exposing the platters, the read/write heads and the spindle shaft.*

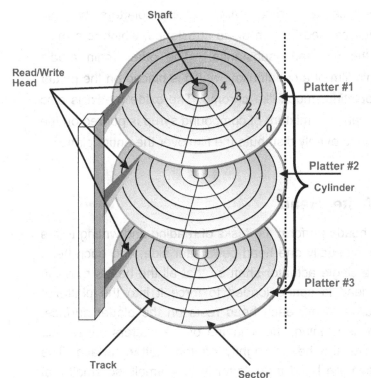

Tracks and Sectors
On the platter at the bottom you can see the labeled portions for tracks and the sector.

Cylinder
A combination of all track #x on each of the platters that spins on the axis of a shaft form a cylinder, where x={0,1,2,..,n}. In the diagram on the left side the combination of tracks #0, on platters 1 to 3 form a cylinder #0. In figure 5.3 there are five cylinders since each platter carries five tracks.

Note:
The Read/Write head read and write data from both sides or surfaces of each platter.

Glass platters can be made much thinner than aluminum ones, and they can better resist the heat produced in operation. The number of platters equal half the number of read/write head a in each Hard Drive Assembly (HDA).

What gives the Platter recording power?

Each platter surface is coated with a film of some magnetically sensitive substance to record data. Some platters become brownish–orange caused by iron oxide substance, which is a main ingredient of the magnetic substance. The other main media consists of a thin film of a cobalt alloy, which is placed on the platter through electroplating, much like chrome. The oxide media is also used to record data. A mixture of compound syrup is poured on the platter, then spin to evenly distribute the film over the entire platter.

The Read/Write Head

The read/write heads perform the tasks of reading and writing to the platters. There is usually one head per platter side, and each head is attached to a single actuator shaft so that all the heads move in unison. Each head is spring loaded to force it into the platter it reads. When drive is off, each head rests on the platter surface. When the drive is running, the spinning of the platters causes air pressure that lifts the heads slightly off the platter surface. The distance between the head and platter is very small, so small that the HDA must be assembled in a clean room as one dust particle can throw the whole thing off. The sensitivity and accuracy involved in HDA assembly is what causes only bigger companies to be able to repair hard drives simply because of the expense of a clean room. A slider is attached to each head. The slider mechanism actually glides over the platter and holds the head at the correct distance to do its job.

The head actuator

All the heads are attached to a device, which is called the head actuator. This part is in charge of moving the heads around the platters. They come in two types: *stepper motor actuators* and *voice coil actuators*.

The stepper motor design is actually an electric motor that moves from one stop position to another, governed by click stop positions. They cannot stop between stop positions. The motor is small and is located outside the HDA, so it is visible from the outside. The stepper motor design is inferior. It suffers from slow access rate and is very sensitive to temperature. It is also sensitive to physical orientation and cannot automatically park the heads in a safe zone. Besides, the actuator operates blindly from the track positions, governed only by the stop positions. Over time, the drive becomes misaligned, requiring occasional re-formats to realign the sector data with the heads.

The voice coil actuator is found in all modern drives, including any over 100MB in capacity. Unlike the stepper design, the heads get feedback about the current position assuring proper tracks are read. The guidance system used by the heads is called a servo. Its job is to position the head over the correct cylinder. It does this through the use of grey code. Grey code is a special binary number system in which any two adjacent numbers provide information to the servo as to their position on the drive. Also, the heads are free to move wherever needed with no steps in between.

When the hard drive is powered down, the springs attached to the heads pull the head into the platter. This is called a *landing*. Every drive is designed to handle thousands of *takeoffs* and *landings*. However, since the head actually hits the platter, it's best to have this happen on a section of platter where there is no data. In a voice coil design, small springs drag the heads into a park and lock position before the drive even stops spinning. This assures that the heads are not uninhibited and left to drag along the platter until the platter stops which is a common problem with the stepper motor design. When powered on, the drive automatically unpark itself and the parking springs are overcome by the magnetic force.

The spindle motor

The spindle motor is responsible for spinning the platters. These devices must be precisely controlled and quiet. They are set to spin the platters at a set rate, ranging from **3600 RPM** to **7200 RPM**. The motor is attached to a feedback loop to make sure it spins at exactly the speed that is required. The speed is not adjustable during operation. Some spindle motors are on the bottom of the drive, below the HDA. The more modern ones are built into the hub of rotation of the platters, thereby taking up no vertical space and allowing more platters.

Attached to the spindle motor is a **ground strap**, which helps rid the drive of the static charges created by rotating the platters through the air. In many drives this can be accessed, by removing the logic board. After a while, this strap can become worn out and produce noise, like a high-pitched squeal. One can usually lubricate the strap and stop the noise, but this process entails some minor disassembling of the drive.

The logic board

The logic board is a preamplifier chip installed underneath the drive. It controls the spindle, head actuator, translates data to a form usable by the controller and the rest of the system. Some logic boards have an integrated controller. Sometimes, an apparent disk failure is actually a failure of the logic board. In such a case, you can replace the logic board and regain access to the data that is in hold status on the drive. This is relatively easy to do because the board is simply plugged into the drive and held in by screws.

Air filter and pressure ventilation.

Air filter and pressure ventilation is used to remove loosened microscopic particles which are discharged through minor wear of internal components and occasional contact of the heads with the platter surface within the hard drive assembly (**HDA**). This is a smart way of preventing the heads from causing damage to the platters. It also allows minor air exchange from outside of the housing. This provides equalization of air pressure so drives can be used in different environments without risk of imploding or exploding.

Mechanism Of Operation Of Hard Disk Drives

Hard disk drives are fragile precision instruments that operate mechanically, so they need to be handled with care. They do not handle shocks very well, and you don't want to replace them. Hard disk are usually very expensive. Laptop computer hard drives are still fragile, but built to handle more shock. Be careful to handle hard drives with extreme care, be sure not to drop hard drives, which can cause severe damage.

As mentioned earlier, the read/write heads are embedded between the several disks or platters of the hard drive. These heads fly around with a hairline distance between then and the platter. The heads usually have a coil of copper wiring in them. When current is passed through the wiring, the surface under it is magnetized, *creating one bit of data.* The direction of the current in the wiring determines the **polarity of the magnetization**, creating either a *0* or a *1*. To read the data again, the drive's electronics sense the polarity changes and determine a *0* or a *1*. This is why computers understand only binary numerals (0,1).

After it reads this data, it needs to figure out what to do with it. It first separates the data from the clock signals that control the timing of other components. The data is looked at by, the disk controller to see if it is what the computer is looking for. If it is, it keeps going. If not, it ignores that bit of data.

The useful data is converted *from serial form*, one bit after another, *to parallel form* so that it will travel optimally over the computer's bus. On IDE and SCSI drives, this is done by, a data separator that is integrated onto the drive. Other designs convert the data in the disk controller.

Hard Disk Drive Storage Mechanics

The size of the platter surface and spinning rate will determine the storage capacity of the Hard Disk Drive. The Hard Disk Drives can therefore achieve large storage capacity by:

- Having more than one Platter. More platters means more storage space.
- Having higher track density (*i.e. many more tracks per inch*)
- Higher data density, more sectors per track, due to higher rotational speed.

The figure below illustrates how data is stored on platter surface.

Figure 5.4. Platters showing sectors/tracks and zoned-bit recording

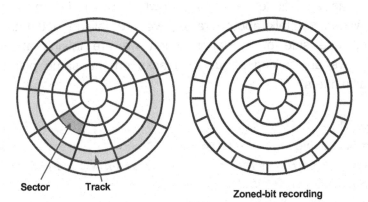

Sector   Track

Zoned-bit recording

Data on a hard drive is stored in *tracks*. These are concentric circles, much like the songs on an ancient LP record. Each track is divided into *sectors*. A *cylinder* is the same track on each **platter**. There are two main ways to record the data on the drive. Modified Frequency Modulation (*MFM*) and Run Length Limited (*RLL*). MFM encoding uses **17 sectors** per track. RLL encoding uses **26 sectors** per track and therefore packages more information in less disk space. With RLL, a high quality surface is needed. If your disk is **RLL** certified by the manufacturer, then its surface is capable of **RLL** encoding. The encoding method is built in to the controller hardware, not the drive itself. If you're using an **MFM** controller with a **RLL** drive, it will work, but your drive is overqualified for the computer.

Multiple Platters or Cylinders

As Hard Disk Drives usually use multiple platters, there are multiple read/write heads inside the Drive. Each disk surface is divided up into circular concentric areas called tracks and each track is divided into a number of sectors. Each sector contains **512** bytes of data.

The tracks are usually referred to as Cylinders because they can be looked at as a Cylinder extending down through the multiple platters. When we talk about Tracks, we are talking about the Tracks on one side of a Platter, when we are talking about Cylinders, we are talking about all Tracks on the same plane, through all platters.

Example 5.1.

The Hard Disk Drive in the first IBM PC/XT Computer had 306 Tracks (Cylinders), four (4) Heads reading four surfaces on two platters, and used 17 Sectors per track. Calculate

a) *the storage capacity of this hard disk drive;*
b) *the total number of cylinders in the volume;*
c) *the total number of tracks per cylinder;*
d) *the total number of sectors in the hard disk;*

* Note: the boot sectors on track 0 is reserved. Exclude this space.

Solution:

a) We assume that the size of a Sector= 512 Bytes

Total storage capacity	= 306 x 17 x 4 x 512 Bytes
Total including the boot sectors	= <u>10,404 Kilobytes</u>.
But net usable storage space	= *305 x 17 x 4 x 512 bytes*
	= 10,618,880 bytes
	= <u>10,370 Kilobytes</u>

b) *The total number of cylinders* = *306 cylinders*

c) *The total number of tracks* = *2 x (2 x 306 tracks per platter)*
= *4 x 306 = 1,224 tracks*

d) *The total number of sectors* = *17 x 1224 = 20,808 sectors*

Clusters

A hard disk is a large string of clusters. Some drives have clusters that consist of 15 sectors. DOS reads and writes from the disk in increments of clusters, and not by sectors. The problem this creates is that you can run out of drive space before you reach the theoretical limit of your drive. Some large drives have clusters that are **32**KB. If you store a file that is **1**Kb, it will take up one **32**KB Cluster. Similarly, if you store a **33**KB file, it will occupy 2 clusters because of the way DOS writes to the disk. Therefore, it is possible to fill up a **1**GB drive with **800** MB of data and not be able to fit another bit into it. The only way to avoid mismanagement of space is to reformat your hard drive and reduce the size of the partitions. This problem is caused by the use of a 16bit FAT in DOS and in previous versions of Windows. In FAT16, the largest partition size that could be used was **2.1**GB and this required a 64 Sector Cluster.

Solution to Cluster problem

This problem is being overcome with the introduction of a **32-bits FAT** in the OSR2 release of Windows 95 and in Windows 98. You can see the cluster size of a Hard Disk Drive by using the CHKDSK command from the DOS prompt, either under DOS 6.xx or under Windows 95 DOS. You can read CHKDSK reports on the *bytes in each allocation unit*, this refers to the cluster size in tables 5.1 to 5.3 below.

Table 5.1. Cluster sizes for FAT16 file system

Partition size (up to)	Number of sectors per Cluster	Cluster size in bytes
128 MB	4 sectors	2,048
256 MB	8 sectors	4,096
512 MB	16 sectors	8,192
1 GB	32 sectors	16,384
2 GB	64 sectors	32,768

Table 5.2. Cluster sizes for FAT32 file system

Partition size (up to)	Number of sectors per Cluster	Cluster size in bytes
256 MB	1 sectors	512
8 GB	8 sectors	4,096
16 GB	16 sectors	8,192
32 GB	32 sectors	16,384
2 TB	64 sectors	32,768

Important: *A hard drive formatted with a FAT32 file system will not be readable under DOS, all older versions of Windows, the original Windows 95, and versions of Windows NT less than version 5.*

Historically DOS versions up to and including DOS 3.3 imposed very restrictive limits on Hard Disk Drive Interfaces. The worst limit involved Hard Disk Drives larger than 32 Megabytes. A drive larger than 32 Megabytes had to be split into sections and no section could be larger than 32 Megabytes. This limitation was overcome in later DOS versions and DOS 6.22 can theoretically address a hard disk drive of up to two tera-bytes (two trillion bytes). The only problem is the 16 bit FAT used by DOS and older Windows Operating Systems, lowers this limit to 2.1 GB.

Here is table 5.3 below describing where the original IDE specified 528 Megabytes limit came from. The maximum capacity limits of IDE Hard Disk Drives was the combination of the lowest values for each parameter across the four layers of communication.

Table 5.3. The maximum capacity limits of IDE Hard Disk Drives

	Limit imposed by the Int13 routine	Limits imposed by the IDE interface	Limit on the original IDE specification
Maximum sectors per track	63	255	63
Maximum number of heads	255	16	16
Maximum number of cylinders	1024	65536	1024
Maximum capacity	8.4GB	136.9GB	528MB

For many years now DOS computers have used translation so the numbers reported to the CMOS setup at the time of installation is quite different to the actual organization of the Hard Disk Drive. Translation can be used to make the organization of the drive fit into the limits imposed by the BIOS, by the Hard Disk Drive Interface and by the Operating System.

Therefore, when you buy a hard drive, the advertised capacity is theoretical. There is no rule about the capacity. When you format the disk, this adds the features necessary to determine how much of the disk it can use. This process determines how much you can store, not the claims by the store. Figure on a 5-10% error. Some companies will boast their "formatted capacity". This measure is a more accurate one.

Storage Capacity Limitations
The hard disk size like any other computer component has some limitations. These limitations are imposed by layers in the system. Disk access involves four layers.   Data must travel through all four layers on its way between the DISK and the user. These layers are
1.   The Hard Disk Drive
2.   The Hard Disk Drive Interface
3.   The BIOS service routines (in particular, Int 13 routine)
4.   The Operating System.

Limits imposed by Hard Drive Interface

The type of hard drive interface may impose size limitations to the hard drive. This imitations are continually being improved as computer technology evolves. For example, the original **ST506 Hard Disk Drive Interface** could only support **8 Heads** but manufacturers did not adhere to the rules and most of the last of the ST506 Hard Disk Drive Interface cards supported **16 heads**. This means the ST506 Interface imposes the first constraint, the disk could have no more than 16 heads.

The newer **ESDI** Hard Disk Drive Interface supported up to 256 heads, up to 4096 Cylinders, and up to 256 Sectors per Track. Given the sector size of 512 bytes, this would mean an ESDI drive could have a capacity of up to **135 GB** in the early 1980s.
The IDE Hard Disk Drive Interface, universal device interface, and SCSI, do not need to know about the numbers of Heads and Cylinders as they are designed quite differently. They are only concerned with the total number of Sectors on the Hard Disk Drive.

Limits imposed by the BIOS

The BIOS is a series of routines, stored in a ROM, that tests, configures and boots up your computer and provide services to the Disk Operating System and application software. BIOS stands for Basic Input/Output Services. In some way the BIOS sets the disk size limits, by responding to those limitations imposed by layers in the system.

The BIOS provides disk services via a software interrupt service routine, **Int 13 hex.**. These services were originally designed for Floppy Disk Drives and this is the cause of many of the limits. Three eight bit registers are used to store the **Track, Head** and **Sector** detail. The **CH** is used to store the *Track number*, the **DH** register stores the *Head number* and the **CL** register is used to store the *Sector number*. Since 8-binary-bits can represent only **256** values, it means that this BIOS service routine can support up to **256 Tracks, 256 Heads, and 256 Sectors per track**. The above scenario represents a capacity of over 8 Gigabytes, not a bad insight for 1982, when Hard Disk Drives were small enough to fit inside a PC. Computers were just becoming a reality. The only problem was the arrangement was wrong, A Drive with more than 16 heads is impractical and even the first Hard Disk Drives had more than 256 cylinders.

This BIOS imposed limitation was overcome by modifying the Int 13 routine for Hard Disk Drives. The use of the CL and CH registers was changed. The low six bits of the CL register stored the number of sectors and the extra two bits not used in CL were tacked onto the front of the 8 bits in the CH register to give 10 bits for the number of cylinders. This means the Int 13 routine, when working with Hard Disk Drives, could recognize a maximum of 1023 Tracks, a maximum of 63 Sectors per track, and a maximum of 256 Heads (sides). The 1023 Track barrier is a problem as modern drives can have as many as 5000 Cylinders.

Hard Disk Drive Increased Storage Capacity

This is basically due to an advance in BIOS to work with drives larger than 504 MB. This limit was there basically because of the geometry in the drive. Newer enhanced BIOS are capable of using translation modes, thereby using different geometry when talking with the drive than when talking with the software. If your BIOS is dated around 1994 or later, it is probably enhanced. You can tell if it offers settings called LBA, ECHS, or Large, which are just three different methods of translation.

Types of Hard Disk Drives

As discussed earlier, the speed and capacity are the most important factors to determine which hard drive is best. You also learned that the speed holds two important measuring features, rotational and interface. Today's computer market may carry a few types of hard disk like **IDE** or **SCSI**("**skuzzy**") that carry these features on them as the most important factor. **IDE** is Integrated Drive Electronics and **SCSI** is Small Computer Systems Interface.

The IDE (Integrated Drive Electronics)

IDE (Integrated Drive Electronics) refers to any drive with the **controller built-in**. Each storage device carries two important features, rotational and interface. The interface most frequently used and misnamed IDE is actually called **ATA** or AT Attachment. The IDE is currently flooding the market with several improved versions, like **EIDE** *(Enhanced IDE)*, **ATA** *(AT Attachment)* and **DMA** *(Direct Memory Access)*. This means that among the IDE types, there may be **non-intelligent** and **intelligent** types.

Non-Intelligent IDE

The first type of IDE was actually Non-Intelligent IDE. Non-intelligent IDE was simple and only responded to the first eight commands built into the original WD1003 controller. Unlike today's drives, most of these drives may be low-level formatted with a few optimizations built on. Sometimes factory defects are written as a file to the drive. This means that, although you can low-level format the drive, it would erase the factory optimizations and defect list. Some companies released programs to do this while saving these settings, but many did not.

Intelligent IDE drives were enhanced to use special commands like the **Identify Drive** command. Intelligent Zoned Recording IDE is an intelligent drive with special Zoned Recording capability. This means that the drive can have a different number of sectors on each track. Since the BIOS can still only handle a fixed number of sectors per track, the drive runs in a special translation mode. This ability means that you cannot low-level format this type of drive without a special program from the manufacturer. These Intelligent IDE's also have a version name of ATA-1 series. The enhanced versions of IDE also fall into the ATA-2 series discussed below.

EIDE (Enhanced IDE)

EIDE is simply Enhanced IDE. EIDE has less of the limitations of the original IDE interface. You can put as many as four devices on one controller. EIDE also allows non-disk devices to be used, such as CD-ROMS. The original IDE allowed only hard drives. EIDE allows the use of much higher capacity drives, up to 9 GB or so. It has a transfer rate of around 11.1 MB per second and above, much faster than IDE, and it also allows you to take advantage of the PCI interface from your new video card.

ATA-2 is EIDE, or Enhanced IDE. This is an extension of the original ATA that includes features such as PIO and DMA modes. PIO stands for Programmed I/O, and DMA stands for Dynamic Memory Access. These are basically performance-enhancing features that include increased capacity, fast data transfer, ATA pocket interface (ATAPI for CD-ROM), and dynamic memory access. These main benefits are discussed below.

Faster Data Transfer

This is a very important feature in today's EIDE drives. ATA-2 offers several different modes for higher performance. Most drives today are capable of PIO Modes 3 and 4, which are very fast. PIO (Programmed I/O) modes determine the speed at which data is transferred to and from the drive.

To run Mode 3 or 4, the IDE port must be on a VL-bus or PCI bus connection. Some newer boards with two IDE connectors only have the IDE 1 connected to the PCI bus, while the second IDE connector uses an ISA bus, only capable of Mode 2. One should look into this before buying a new motherboard. Below is a table of the PIO Modes:

Table 5.3 Table of Programmed I/O Modes and Transfer Rates

PIO Mode	Transfer	ATA Version
0	3.3 MB/sec	ATA-1
1	5.2 MB/sec	ATA-1
2	8.3 MB/sec	ATA-1
3	11.1 MB/sec	ATA-2
4	16.6 MB/sec	ATA-2

ATAPI.

This is ATA Packet Interface. This is designed for extra drives like CD-ROM's and tape drives that connect to an ATA connector.

DMA Transfer.

ATA-2 drives support Direct Memory Access transfers which means that data is transferred directly from the drive to memory, bypassing the CPU. Only ATA-2 drives support this feature. Most operating systems(OS) do not support it. Proper software support for this transfer is needed before it can be taken full advantage of.

SCSI Drives

SCSI (pronounced **"skuzzy"**) is Small Computer Systems Interface. SCSI drives are independent. **SCSI** drive bus is completely separate from the usual buses, such as PCI or ISA. It does not rely on the BIOS to be able to talk to the computer. SCSI drive bus is completely independent.

When the computer boots, it checks for additional hardware ROMs. As it does this, it finds your SCSI adapter card, if you have one. It gets no details as to what is attached to the adapter. You can connect as many as seven SCSI devices to the adapter without the computer's knowledge. The adapter keeps track of the data flow across the SCSI bus. Each device gets its own SCSI address. Each device can talk with the other SCSI devices across the bus, all independent from the computer.

The SCSI interface certainly speeds up the computer, but many problems come with the added speed. There are no set standards for SCSI; there is the original SCSI and the newer SCSI-2. These two sometimes have a hard time talking to each other in your computer. Also, many companies have developed proprietary SCSI standards. A drive like this may not be able to talk with a SCSI device of another make. SCSI hardware requires drivers unique to the operating system you are using and your special hardware combination. Each device requires its own ID number, with the end of the chain being set to be the end device. For these reasons, configuring SCSI is considered hard.

Hard Drive Interface

For data stored in the storage device to be useful, the hard drive should be able to communicate with the CPU to exchange data. The of device that allow this communication is known as the drive interface. When you open up your computer and examine your drives, you will see the kind you have. SCSI, IDE, and EIDE each have a single ribbon running from the drive to the adapter card.

SCSI ribbons contain **50 wires**. IDE ribbons have **40**. Both cables are a straight design, no twists. A floppy ribbon has a small twist of seven wires in the ribbon. Also view the diagrams below:

Figure 5.5. The IDE 40-wires wide ribbon cable

Figure 5.6. The IDE 40-Conductor Cable and 80-Conductor Cable

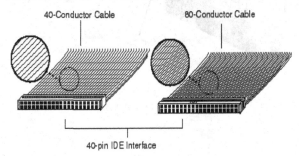

ATA I/O

The ATA interface uses a 40-pin connector. This is usually designed to prevent plugging it in backwards -- this design is recommended. Plugging it in backwards can damage the drive and related circuitry.

The ribbon cable is 40-wires wide. It carries all signals to and from the controller. This cable should be no longer than 18 inches long.

For a connector pin-out and discussion of important signals, see ATA Connector Pin-Out.

Figure 5.7. Installation of IDE Hard drives

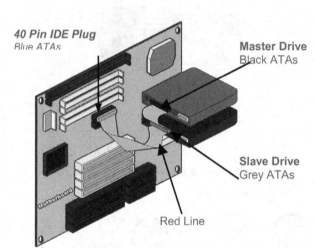

Life Expectancy

Hard disks will not last forever. They are mechanical devices with motors. Newer drives are much better designed than the old ones. They are sturdier, more reliable, use less power and are less likely to crash. However, hard disks will not last forever, so plan to replace after a period of time.

A typical user who takes good care of his/her system should get many years of service from a hard drive. It is possible, though, for it to crash in a few weeks.

In most cases, the data on your computer is the most valuable thing about your computer. For this reason, it is best to act as though your hard drive can crash at any time.

Floppy Disks and Drives

Floppy Disks

Floppy disks are portable storage systems commonly used in PC systems. Like the hard disk, we can identify 3.5-inch (3½") and 5.25-inch (5¼") as the two common sizes of floppy disks used in PC systems. The diskette has a hard plastic shell case that protects the disk internal magnetically stored data. Formerly diskettes were of 5.25-inch size having as little capacities as 360 KB or 720 KB.

The most popular size is the **3.5-inch diskette**— A standard type of storage medium used in most PCs that works with a corresponding 3.5-inch diskette drive. These diskettes come in different *densities*. The old ones are single-sided and only hold 720 KB of data. The standard disks of today are called high-density. They usually hold 1.44 MB of data. The newest disks are called extended-capacity and hold 2.88 MB. Like the hard disk, before a floppy diskette can be used, it must be formatted. Formatting prepares the disk so that the drive can use it. To do this, type FORMAT A: at the DOS prompt

(C:), When Windows asks for a label, it wants to know what you want to call the disk.

These portable diskettes are flat 3.5-inch squares as shown in the figure below.

Figure 5.8. Five 3½" floppy disks formatted to 1.44MB

Floppy drives

Floppy drives are the slots on the front of your computer that allows you to insert disks, copy files to, and install programs from. Old ones are 5¼" wide which are significantly bigger. They accept the big 5¼" disks that can't hold very much data. Almost all of us now have 3½" drives. These drives accept 3½" disks. Like the disks floppy drives come in high-density or low-density. So the density of the disk must always match with the density of the floppy drive. This means you cannot use a high-density disk in a low-density drive, but the vice versa may work alright.

There are two type of floppy drives: internal and external. The figure below shows the two types.

Figure 5.9. Floppy drives. The first figure on your left is an internal floppy drive while the other is an external floppy drive .

> Notes: You cannot use a high-density disk in a low-density drive. Also, when formatting, make sure you type A:. Do not type any other drive, especially C:, otherwise you're in for trouble. Floppies are almost always A:. Also, disks can go bad. If a particular diskette has many errors, trash it.

Structure Of Floppy Drive

The inside of a standard floppy drive has many similarities to the inside of a hard drive. It has the same basic parts. Most floppy drives have two read/write heads, meaning that it is double-sided. These heads are used to read and write data to the diskette. Like a hard drive, the head mechanism is moved by the head actuator. It is a stepper motor design, much like the old hard drives. This small motor moves the heads in and out, giving them the ability to position themselves over any track on the disk. The heads use a recording method known as **tunnel erasure**. Basically, this recording method is a way to keep track of data. As the head writes the data to a track, the tunnel erase heads come along and erase the outer edges of the track, thereby creating a sharp-edged track, very distinct from the others. This process keeps the data on one track from being confused with data from another track, thereby eliminating problems.

The heads are spring loaded. Therefore, they are physically contacting the disk while they are reading and writing. The drive spins the disk at about **300 RPM**; therefore, this contact is not a problem to the data because friction is minimal. Some diskette manufacturers coat the disk with Teflon, further reducing friction between the heads and the disk. Eventually, a build-up of Teflon will form on the heads, requiring you to clean them.

The spindle motor spins the disk. As said before, it spins the disk at **300 RPM**. On old 5.25" drives, the disks used to spin at 360 RPM. Older drives had the spindle motor attached to a belt system that spins the disks. These were not very reliable and the RPM was not constant. Today's drives use a direct-drive system with no belts. This system has automatic torque compensation so that sticky disks are made to spin with greater force than a slippery one, thereby maintaining a constant **300 RPM** with all disks.

The newer drives have this automatic ability, while older ones require periodic adjusting. This is done by using the strobe marks which are on the motor. You run the drive in fluorescent light and adjust the RPM until the marks look still, just like a car rim or wagon wheel when turned fast enough.

Underneath the drive, you have the logic board. Like the hard drive, it serves to control the internal parts of the drive and serves as an interface between these parts and the floppy drive controller. All floppy drives use the SA-400 interface. For this reason, any floppy drive will work with any computer, right out of the box.

The front of the drive is called the faceplate. This is the part of the drive which is visible from the front of the case. These faceplates come in several sizes and colors. Some faceplates are larger than the rest of the drive, requiring you to install the drive from the front, a habit you'll probably form anyway.

The connectors are on the back of the drive. There are two: the power connector and the ribbon cable connector. The only difference between the power and ribbon connect is that the large 5.25" drives use a larger power plug, similar to the type used in a hard drive. If you are installing a 3.5" into an older case, you might not have the small-type connector available. In this case, you will have to buy an adapter to convert the large plug into the small type. The 3.5" drives use a smaller plug. All floppy drives use the same **34-pin** data cable.

Occasionally, you might find a floppy drive that has weird connections. Some drives have one 40-pin connector that carries both the power and the data. Others use one 34-pin cable that carries both. However, these drives are rather rare.

External Drives, Cartridge-based Storage
There also are other types of storage options for PCs, such as LS-120 Zip, and Jaz drives that are appearing as standard components

in some systems. These are cartridge-based devices that increase the PC's storage capacity and can travel easily from one computer to another. The use of these storage devices may depend on the users need for storage space. If you plan to design photo albums or store a lot of company graphics, the more storage you have available, the more flexibility. Items that will easily eat up your valuable storage space include video clips, large amounts of software installed on the system, and photo archives. In the above, case systems that feature cartridge-based options or external drive may be needed. However, keep in mind the speed of external drives is generally slower than with an internal drive. If you're planning on conducting any multimedia activities, such as sound or graphics, plan on cartridge options that store at least 100MB.

The Zip drive is a type of storage system designed by Iomega Corporation that holds up to 100MB of data on portable diskettes. Zip drives are popular devices used for storing, transporting, and backing up files. The figures below are examples of ZIP drive and disk.

Figure 5.10a

The Zip drive

Figure 5.10b
Zip Disk

The zip drive has many similarities to the floppy drive. It is the most preferred portable disk since it can hold much more data than the floppy drives. Like the floppy drive the zip drives come in two types,

---

internal and external. The internal drives are faster than external drives but the latter may be more flexible since it can be easily unplugged and used directly in another system.

Figure 5.11a Internal Zip Drive   Figure 5.11b External Zip Drive

Compact Disc Read Only Memory or CD-ROM

It is a permanent storage device that has a minimum storage capacity of 500 MB to 680MB. It is very useful when it comes to storing large files on removable disk. Since CD-ROM drive is read only we cannot write data in that drive. To be able to put data on CD the system must be equipped with a compact disc-recordable (CD-R) or compact disc-rewriteable (CD-RW) drive. This may be a better choice than a Zip. The design concept has a close similarity to that of floppy drive and disk. The only difference may lie on the CD-ROM disk itself, which comes without the protective plastic shell, and may also required a special device and software before your system can write to it.

CDROM drive can be mounted Internally (inside the PC) or in an external enclosure. The first CDROM drives used a Caddy that protected and supported the CDROM but now all drives are "Caddy Less". The current CDROM technology provides 650 to 680MB of storage space at a fraction of magnetic media's cost per megabyte. CDROM is now the most cost-effective distribution medium. Most software packages and operating systems are supplied as standard, on a CDROM.

CD-ROM Drive Interfaces

A CDROM drive requires an interface circuit to interface it to the computers bus and this is often provided on the Sound Card. Until recently CDROM drives used either a SCSI interface or a proprietary interface, based loosely on the ISA bus. This is often called the AT interface.

Figure 4.9 Internal CD –ROM Drives

CD – ROM Drive Speed/Performance Specifications

The CD-ROM performance can be measured in terms of the drive's seeking or access time, that is how long it takes the drive to find a piece of data on the disc and its data-transfer or throughput rate. Typical performance figures for CDROM drives are, access time **280msec**, burst data transfer rate of 2.5Mbyte/sec in async mode and 4.0Mbyte in sync mode and hard error rates of better than 10 to the power of 12 (after error correction).

Historically the first CD-ROM Drives had a data-transfer rate of **150 Kbytes per second** but it was not long before "MultiSpin" Double Speed drives provided 300 Kbytes per second. Triple speed drives had a very short life span and were replaced by Quad speed drives, providing 600 Kbytes per second data-transfer speed. In the middle of 1996, Hex and Eight speed drives were the industry standard with data-transfer rates of 900 and 1200 Kbytes per second. Early in 1997 the 12 speed drive was the norm followed by 24 and 32 and now 48 speed drives are the state of the art. We however see the speeds written in terms of Xs starting with 1X = 150 Kb/sec, 2X = 300 Kb/sec, and 32X = 4800Kb/sec (32x150Kb/sec).

When a manufacturer talks about a 24 speed or 32 speed he does not usually mean what he did when talking about two and four speed drives. Today the speed rating is the maximum possible when reading the outer most tracks. When reading the inner most tracks a sixteen speed drive is no faster than an eight speed drive. If the system has a CD-ROM drive, it should be no slower than 24X or 32X, which is the manufacturer standard. CD-RW drives, which allow you to read and write to a CD-RW just like a hard drive, also are appearing on new systems. CD-RW drives presently write at 1X, 2X, and 4X, and play back regular CD-ROMs between 10X and 24X. CD-ROM drives are faster than today's CD-RW drives, especially when it comes to installing software, but a CD-RW drive is handy for recording data. CDROM Drives, like Hard Drives, get a performance boost from caching. Some CDROM Drives have as much as 256K of built-in buffer memory.

DVD-ROM

Modern computers have faster multimedia CD drives also known as Digital Video Disc drive or DVD-ROM. The DVD-ROMs are the present rage because they can read audio CDs, data CDs, and DVD-ROMs. The advantage of a system with DVD is the versatility it offers as manufacturers begin to use this medium more. The average speed of a DVD drive is 2X to 6X, which is about equivalent to 18X and 40X in CD-ROM speed.

Figure 4.10 Internal CD –ROM Drives

This is similar to the CD-ROM drive except that the DVD performs under higher speeds

ZipCD

The External USB Iomega® **ZipCD** drive is the perfect complement to any hard drive. It reads, writes, or rewrites many types of CD formats, making it an ideal storage solution for users who want to copy and share files with co-workers, service bureaus, and even friends and family. The External USB ZipCD drive performs at a speed of 4X Record, 4X Write, and 6X Max Read. You can use standard CD-Recordable (CD-R) and CD-ReWritable (CD-RW) discs to organize and archive files without cluttering up your hard drive or the network. And for road warriors, CDs are an efficient way to transport and exchange large files. It can be used to back up and archive important files, or share multimedia presentations, photos, and Internet downloads. The ZipCD drive allows you to copy your original music on CDs and play them in your home or car stereo. With bundled software to enhance performance, the ZipCD drive is a complete solution.

BACKUP TAPE DRIVES

The figure 4.11 below is an external tape drive. It is a popular device commonly used to back up data. The backup tape operates as a sequential read/write device. This means that once the tape is Initiated it has to complete the

process without interruption. Figure 4.11 External Tape Drive

Transferred to any of the mentioned storage devices before you can use it. In the market today, you can find several media types of backup tapes, that may include Exabyte Mamoth, Quantum DLT (Digital Linear Tape), HP, IMB and Seagate LTO's (Linear Tape Open), and Sony AIT tapes(Advanced Intelligent Tape). Digital Linear Tapes, DLT is the top of the line tape backup option and is designed to backup large amounts of data in a short amount of time. DLT offers high transfer rates (5 MB/sec) and large storage capacity (10-70 GB per tape). DLT can be arrayed to back up a very large network file system.

Tape Speed and Stress

Linear drives move tape at a relatively fast rate, typically over **150 inches per second** (**ips**). The helical scan drives use a much slower tape speed of less than one ips through the tape path and past the rapidly rotating drum assembly. Interestingly, the relative tape speed is nearly equal in both helical-scan and linear technologies. Modern tapes have width ranging from 4mm to 8mm with data transfer rates between 1MB/sec and 16 MB/sec. Tapes are slower, they usually have media load and file access time ranging between 10 seconds and 55 seconds.

Exercise 5

Students are required to select from the five alternatives labeled A-E the most appropriate answer to the questions 1- 25

1. Which of the following processes will prepare the storage device to store data by organizing into tracks, sectors and clusters?
   A. Optimization
   B. Partitioning
   C. Formatting
   D. Tunneling
   E. All of the above

2. During disk partitioning the storage device such as the hard disk is divided measurable areas known as?
   A. Clusters
   B. Sectors
   C. Tracks
   D. Cylinders
   E. Volumes

3. Which of the following is **NOT True** about storage devices?
   A. The connecting Bus has rotational speed.
   B. They all have rotational speed.
   C. Their platters have transfer speed.
   D. They have interface speed measured in MHz
   E. They have both rotational and interface speeds

4. What's the main purpose of hard drive in Computer systems?
   A. It stores data permanently for the computer.
   B. It's part of the CPU hardware.
   C. It implements the virtual memory in the CPU.
   D. It contains a collection of chips that process data.
   E. It's storage tank for the CPU during program execution.

5. Which of the following is true about Dynamic Memory Access hard drive?

    A. It's the computer's main memory

    B. It accesses CPU data dynamically

    C. By-passes CPU to access data in RAM directly

    D. It's volatile.

    E. A, B, and C.

6. The computer memory which stores information indefinitely and that can be easily changed by users as needed is popularly known as ?

    A. Hard drive

    B. RAM

    C. Cache

    D. ROM

    E. All of the above

7. In PCs there are two groups of memory devices, primary and secondary. Which of the following is classified as secondary?

    A. Virtual Memory

    B. Cache

    C. Hard Drive

    D. CD-ROM

    E. C and D only

8. Which of the following may apply to the spinning speed of your PC hard drive?

    A. 720 MB/Sec   D. 66MHz

    B. 7200 rpm   E. 16 MB/Sec

    C. UDMA-33

9. In MS Windows operating system the maximum number of hard disk partitions you can possibly create per Computer is.

   A. up to 4 partitions   D. up to 24 partitions

   B. up to 20 partitions   E. up to 26 partitions

   C. up to 22 partitions

10. When selecting your PC storage device which of the following will be the most appropriate order of importance.

    i. Speed   ii. Capacity   iii. Connectors

    A. i, ii, and iii   D. ii, iii, and i

    B. ii, i, and iii   E. iii, i, and ii

    C. i, iii, and ii

11. In the modern PC world there are two major groups of hard drives are?

    A. IDE and EIDE only

    B. IDE and ATA-X only

    C. IDE and SCSI only

    D. DMA and EIDE only

    E. None of the above

12. Assuming each cluster size of the hard disk of your Pentium PC is 32 MB. How much disk space would be required to store a file size of 33 MB?

    A. 32MB   D. 64MB

    B. 33MB   E. 66MB

    C. 40MB

13. The files of the operating system such as DOS or Windows running on your PC are stored on your hard drive in multiples of

   A. Tracks

   B. Sectors

   C. Clusters

   D. FAT

   E. Cylinders

14. The same track on each platter of a hard drive constitutes what is popularly known as

   A. Partition

   B. Volume

   C. Cylinder

   D. Cluster

   E. Sector

15. The hard drive has a capability of storing data thanks to

   A. Magnetically sensitive substance

   B. Aluminium alloy

   C. Ceramic platters

   D. Glass fibers

   E. Cobalt

16. Different operating systems allow different partition size on hard drives. Assuming your PC is running DOS 7.0 what is the allowable maximum partition size that can be created?

   A. 2 bytes   D. 2 gigabytes

   B. 2 kilobytes   E. 2 terabytes

   C. 2 megabytes

17. The major function of the hard disk drives logical board is

    A. to store data permanently

    B. to control the spindle of hard disk drive

    C. to control the head actuator of hard disk drive

    D. to translate data to usable form

    E. None of the above

18. What does polarity of magnetization produce in hard drives?

    A. Converts current to 0 and 1 and vice versa

    B. Attracts data using magnetic force

    C. Converts Hexadecimal figures to binary

    D. Controls disks electricity and magnetic force

    E. Determines the data flow sequence.

19. Which of the following storage devices does **NOT** use the surface area of platter(s) for storing data?

    A. CD-ROM   D. DVD-ROM

    B. Floppy disk   E. UDMA-33 Hard Disk

    C. Backup Tape

20. Which of the following facts is/are true about a floppy disks

    A. 3.5 '' is bigger than 5.25" in terms of storage capacity

    B. 5.25" disk is always faster than 3.5" disk

    C. Low density disk will not work in high density drive

    D. A and C only

    E. all of the above

21. Which of the following is the main cause for your computer to freeze?
    A. insufficient memory – RAM
    B. thrashing
    C. insufficient storage cells on the hard-drive
    D. all of the above
    E. only A and C

22. The process of using part of permanent storage RAM is
    A. virtual memory   D. Physical Ram
    B. Cache memory   E. EPROM
    C. Tunelling

23. The process by which to make maximum use of the system resources available for efficiency is known as:
    A. Upgrading   D. Caching
    B. Formatting   E. Optimization
    C. Installation

24. The purpose of virtual memory is to:
    A. Upgrade the storage capacity of computer.
    B. Enlarge the address space of the system.
    C. Double the cache memory space CPU.
    D. Double the cache memory.
    E. Manipulate the hard drive.

25. What is maximum size of virtual address can a 32-bit computer machine running under windows 95/98 generate
    A. 25 kilobytes   D. 8 Gigabytes
    B. 128 megabytes   E. 32 Gigabytes
    C. 4 Gigabytes

The Computer Peripherals   Chapter 6

- o   Overview of Peripherals
- o   Expansion Slots and Internal Peripherals (video & sound cards)
- o   External Peripherals Devices:
  - o   Computer Monitor
  - o   Keyboard and Mouse
  - o   Printer and Scanner, Modem

Overview of Peripherals

In the previous chapters we briefly described computer hardware. We now know that the motherboard, memory, processor, monitor, keyboard, mouse, modem, video card, sound card, hard drive and all other nuts and bolts stuff are known as hardware. The figures on page 44 are all hardware. Which hardware components are said to be peripherals? Well, the word *peripheral* simply means auxiliary or supplementary parts. In the PC world peripheral refers to any device connected to a computer to provide communication or auxiliary functions. The communication may involve input and output, while auxiliary function may imply additional storage. So, wait a minute! Don't be confused when separating peripheral components from the main computer unit. When we take the CPU box or the system unit which is shown in figure 6.1, the remaining components are all seen as auxiliaries, since they also play an important role in the computer, they are known as peripherals.

Figure 6.1 An illustration of the system unit(CPU) its peripherals.

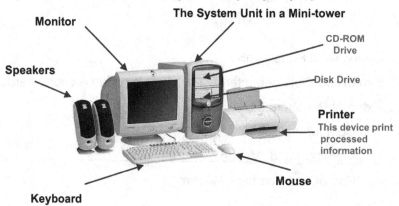

The System Unit in a Mini-tower

Monitor

CD-ROM Drive

Speakers

Disk Drive

Printer
This device print processed information

Mouse

Keyboard

In the figure 6.1, all the peripherals are exposed. These peripherals are separate units other than the CPU and they are classified as **external peripherals**. There are other peripherals hidden in the CPU box. Such peripherals as hard drive, CD-drives, and sound cards are attached internally in the CPU box.

As their name suggests, peripheral devices lie on the edges of the computer system unit. Their job is to provide a connection, or *interface* (*gateway*), to the world outside the computer system (System unit). Peripheral devices in general may include keyboards, mice, monitors, printers, scanners and modems. How about the memory chips, the sound cards, the video cards, the hard drives, the CD-drives, floppy drives and all other components that the CPU box might be housing. Are they peripherals? The answer is yes, in that, without any of these components the CPU will still be a computer, and the CPU will still function as a computer, but the vice versa might not be possible. In other words the CPU plays the same role as the human brain and heart. When the brain stops functioning life ceases to go on even though the rest of the body organs might still be functional. These internal components may be classified as the **internal peripherals**. See figure 6.2 below.

INTERNAL PERIPHERALS

Figure 6.2 shows peripherals which cannot be seen from the outside, or surrounding the system unit box, for the simple reason being that they might be kept inside the system unit box. Watch out!. Practically, every external peripheral must have an *internal device* (*internal peripheral*) installed in your computer that can provide a sort of interface between the CPU unit and the external peripheral device itself. These peripherals that are fixed internally in your computer box, and may have connecting ends as illustrated in figure 6.3. These connecting ends, popularly known as *ports,* serve as gateways through which the CPU is able to communicate with the corresponding external peripherals. The video card, for instance, via its port may connect to the monitor, while a sound card via its port may connect to the external speakers, microphones, etc. The peripherals like the floppy disk drive and CD-ROM drives are usually attached to the CPU box inside   the front as shown in figure 6.2 below.

Figure 6.2 The inside look of a system unit box and the front view.

The connecting ports for most internal peripherals to the external peripherals e.g. video card to monitor, the sound card to the speakers, etc.

The front view may look like this

bays for disk drive

Removable Disk Drive space

Actually, the use of modern computers is now full of fun and flexible, thanks to these peripherals. The peripherals like the keyboard, mouse, monitor and sound blasters make personal computers more friendly to its users. In the subsequent chapters we will discuss each of these peripherals more closely. As you read you will learn more about each of these peripheral components beginning with those closest to the CPU, and ending with the distant components. Under the internal peripherals this chapter will walk you through the memory and storage devices, circuit boards and cards, like video and sound cards.

Expansion Cards

Expansion cards are circuit boards that fit into expansion slots on the motherboard to provide the computer with new devices, such as a graphics card, modem, and scanner. To be able to attach your computer to any of the external peripheral devices your computer must be equipped with the external ports. Each expansion card usually come with two important edges, one part that fits unto the obviously carrying matching features for ISA, PCI, or AGP interface and the other parts, also popularly known as a connecting **port** that provides connections to the external peripheral devices. With the two connecting edges, expansion cards play an intermediary role between the CPU and the external peripherals.

Figure 6.3 below illustrates some popular connecting ports for peripheral devices, which any average computer user may be familiar with.

Figure 6.3 The Connecting ports for peripheral devices

USB Port   Audio Port   Vol. Control   Game Port   Firewire Port

Extension Cable included

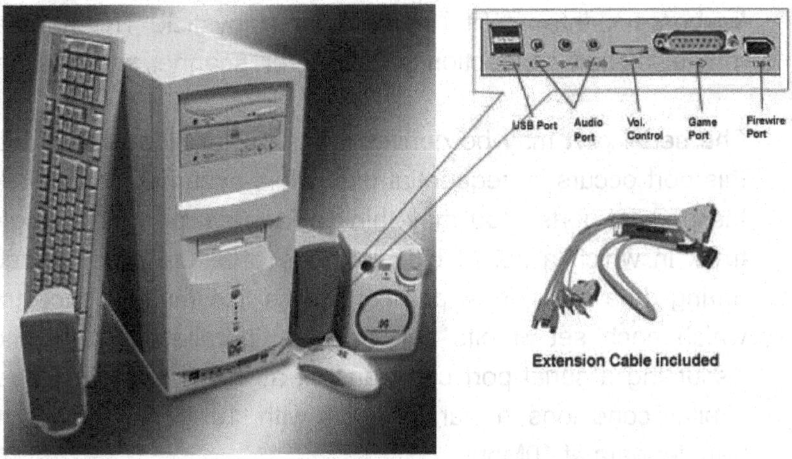

Connecting Ports

The actual communication between the external peripherals and any expansion card slotted into PCI, AGP, or ISA interface may occur through the connecting **port**. Modern PC may be equipped with several types of ports commonly described as **Serial, Parallel, PS/2 and USB** (*universal serial bus)* **connections** also known as **Firewire, and Infra-Red(IR).**

Serial and Parallel Ports:

Often times the difference between the serial ports and the parallel ports may be very confusing. However they come as the most standard gateways to the external peripherals. There are three important attributes that can best describe a PC connecting port. These include the mode of transmission, number of connecting pins and gender of connectors or connecting ends. The mode of transmission could be **serial** or **parallel**. I will use the type of peripheral device to determine the kind of port. For example, your PC may come with several serial ports each allowing you to connect to one of the appropriate peripheral devices like the monitor, mouse, keyboard, printer, digital camera, or a joystick. The parallel ports may also allow connection to the printer, scanner and external drive.

The serial port may be derived from the fact that data transfer via this port occurs in sequential bits, while occurring in parallel bits in the parallel ports. You may think of a serial port as a single race-track in which a set of bits flow one after the other in sequence during data transfer. A parallel port is like multiple race tracks in which each set of bits can travel in parallel at the same time. Assuming a serial port can transmit at the speed of 1Mbps under similar conditions a parallel port with 10 transmitting pins may transfer data at 10Mbps.

Disadvantages of using serial and parallel ports:

The serial and parallel connecting ports may also have some disadvantages. The serial and parallel ports are usually slow. Also changing or attaching a new device on your computer already running often requires shutting down the system completely. The creation of PS/2 and USB ports has added greater flexibility by supporting multiple connections simultaneously.

PS/2 And USB Ports

Nowadays, most personal computers come with PS/2 and USB connecting ports. The PS/2 ports usually connect the keyboard and the mouse. As its name implies the USB port known as the Universal Serial Bus (USB) is a standard that was developed in 1996 as a better solution to handling peripheral-to-system connection. These ports make adding a hardware peripheral, such as a printer, scanner, or joystick, a snap. In fact, Microsoft designed its Windows 98 operating system specifically with USB devices in mind; you simply plug the peripheral into the USB slot, and when you restart the computer, or while the computer is still on, the device should be immediately detected. Gone are the days when you had to worry about configuring hardware or manually adding product drivers. Before you purchase your next system, make sure you have at least two or more USB ports available. The USB technology supports connection of multiple devices simultaneously. The flexibility it provides include:

- o the support for easy plug-and-play installation. Users do not need to configure or set system switches or manually install drivers for attached devices.
- o support for both low and high speed devices which may range between 1.2 and 12 Mbps

- o support hot swapping making it possible to remove and connect peripherals without shutting down your computer
- o support power supply to low-power devices
- o and finally support up to 127 devices per port, through a hub, which is a port extender of sorts.

Connecting Pins

The next thing to watch out for would be the number of connecting pins. A port can also have a gender that has either male or female connectors as shown in figure 6.4 below. It is important to note that a female port can also be parallel depending on which peripheral device it may be connected to. Figures 6.5 & 6.6, illustrates several different types of ports more closely. The analogy of serial or parallel may be due to how these ports communicate with their attached devices.

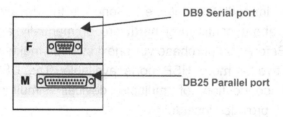

The three categories of connecting ports may be illustrated in the figure 6.5. below.

Figure 6.5. Example of some popular serial ports

4 MiniDIN Female	Used as the keyboard and mouse port on Macintosh computers
5 DIN Female and Male	Used as the keyboard port on IBM PC, XT, AT, and compatible computers.
6 MiniDIN Female and Male	Used as the keyboard and mouse port on IBM PS/2 and compatible computers.
8 MiniDIN Female	Used as the communication port on Macintosh computers. Can be used to connect printers, modems, Appletalk networks, etc.
DB9 Male	Used as the serial port connector on IBM AT, PS/2 and compatible computers for connecting modems and other RS-232 serial devices.
High Density DB15 Female	Used as the VGA or SVGA video connector on IBM, AT, PS/2 and compatible computers
DB15 Female	Used as a joystick port on IBM, AT, PS/2 and compatible computers. Used as the video connector on most Macintosh computers.
DB25 Female	Used as the parallel port connector on IBM, AT, PS/2 and compatible computers. Used as the SCSI connector on most Macintosh computers.
DB25 Male	Used as the serial port connector on older IBM, AT, PS/2 and compatible computers. For attaching modems and RS-232 serial peripherals.

Figure 6.5. Example of some popular serial ports (continued)

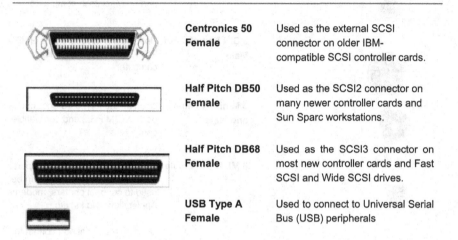

	Centronics 50 Female	Used as the external SCSI connector on older IBM-compatible SCSI controller cards.
	Half Pitch DB50 Female	Used as the SCSI2 connector on many newer controller cards and Sun Sparc workstations.
	Half Pitch DB68 Female	Used as the SCSI3 connector on most new controller cards and Fast SCSI and Wide SCSI drives.
	USB Type A Female	Used to connect to Universal Serial Bus (USB) peripherals

Some Popular Parallel ports Connectors

Visually the parallel ports are not different from the serial ports, but technically they may be different in the way they transmit data. The parallel port transmits data in parallel bits. In the figure 6.6 below you will see a few examples of parallel port connectors. The port connectors from the computer and its peripheral may be designed differently. The figures below show the parallel port connectors viewing from the back of the computer and the printer. You may notice that parallel ports have basic visual similarities to the serial ports.

Figure 6.6. Example of some popular parallel ports

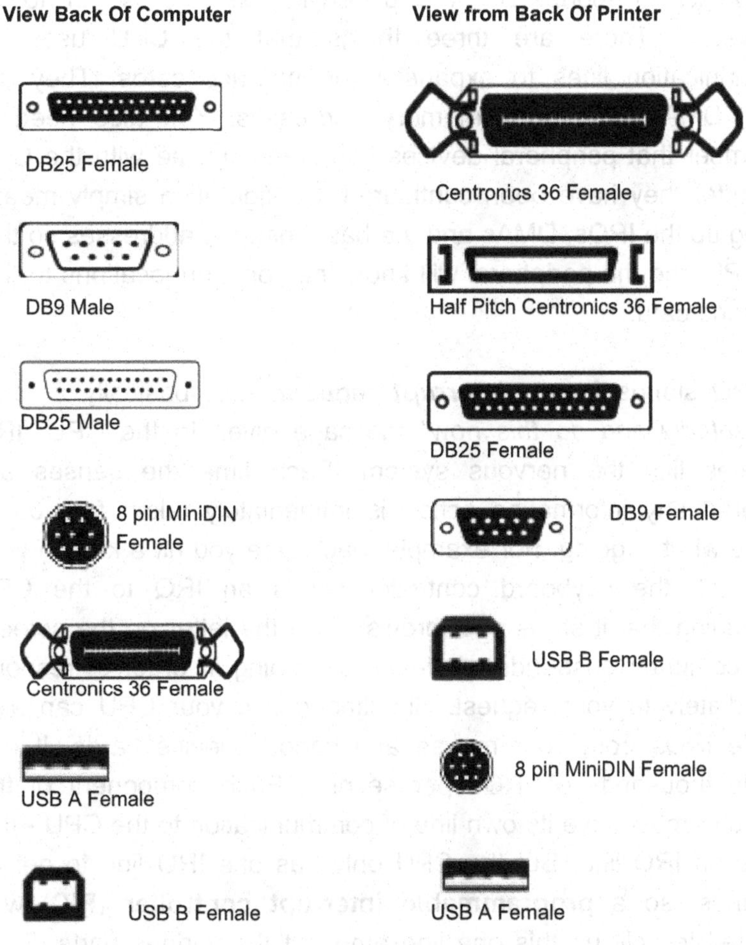

View Back Of Computer   **View from Back Of Printer**

DB25 Female

Centronics 36 Female

DB9 Male

Half Pitch Centronics 36 Female

DB25 Male

DB25 Female

8 pin MiniDIN
Female

DB9 Female

Centronics 36 Female

USB B Female

USB A Female

8 pin MiniDIN Female

USB B Female

USB A Female

Communications between CPU And Peripherals

So far you have learned about expansion slots, cards and communication ports. But the underlying secrete is yet to be uncovered. There are three things that the CPU uses as communication lines to expansion or interface cards. They are IRQs, DMAs, and base memory addresses. You also need to remember that peripheral devices can communicate with the CPU only after they have been configured. Configuration simply means setting up the IRQs, DMAs and the base memory addresses so that the CPU and the peripheral will know the common locations to find and send data.

An **IRQ** stands for an **interrupt request**. It is basically a "*stop momentarily and do this now*" message given to the CPU. IRQ operates like the nervous system. Each time the senses are triggered any information action is immediately taken for you to realize what is going. For example, each time you hit a key on your keyboard, the keyboard controller sends an IRQ to the CPU demanding that it stops and throws down the letter on the screen. Your computer suspends whatever it is doing in order to respond immediately to your request. Simultaneously, your CPU can also handle IRQs from your mouse and various device cards. It can handle thousands of IRQs per second. Each component of the computer must have its own line of communication to the CPU - this line is an IRQ line. But the CPU only has one IRQ line to get its demands, so a **programmable interrupt controller** (**PIC**) was invented to split up this one line amongst the various parts. Each one can handle 8 lines, so we link two together to give the typical amount of IRQs in today's PCs. There are 14 usable IRQs plus two reserved ones. In table 6.1 below you can count 16 IRQs, from IRQ0 thru IRQ15. Each part must have its own IRQ line, so if you set two parts to the same line, your computer will crash or the system will display resource conflict errors. A resource conflict will keep you from using the device cards or just refuse to boot up until you take the parts out, or you take the appropriate steps to resolve the conflict. This one concept is the basis of many heartaches involved with putting device cards in computers. When your CPU responds to an IRQ, it usually sends data back to the part that asked for it.

DMA, or Direct Memory Access

Often, an interface card will ask for the same data again and again, so to save the CPU the time of rethinking the same information and sending it all over, DMA, or direct memory access, was invented. DMA is usually composed of a message sent to the card saying where to look in the memory for specific data. This makes things faster. But once again, these can conflict. Since DMA reserves a little section of memory for the interface cards, one card's part cannot overlap onto another card's territory. Fortunately, not many cards can even use DMAs, so there are rarely such conflicts.

Base memory addresses

Base memory addresses are sometimes called I/O ports or port addresses. They are memory addresses architecturally reserved and assigned to the most fundamental or basic functions such as keyboard input, printer output, network input/output, etc., that require IRQ or Interrupt request. Think of a base memory address phenomenon as the postal box and a drop box. Each house address has a mail box where mails can be received in and a drop box where mails can be sent out or posted. This example illustrates how CPUs respond to IRQs. CPUs can't respond down the same line that it is *getting* IRQs from. So, a different route is set up for the responses. It serves as a link allowing the CPU and the components to talk directly. The port addresses usually in hexadecimal numerals look something like 01F6, or 02F6. For the typical layout of IRQs and DMAs in a computer, the table below shows most common configuration addresses.

Table 6.1. IRQ No. And The Responding Hardware Components

Interrupt No.	Hardware Component That
IRQ0	System Timer
IRQ1	Keyboard
IRQ2	Some video cards
IRQ3	COM2, COM4
IRQ4	COM1, COM3
IRQ5	Sound Card
IRQ6	Floppy drive controller
IRQ7	LPT1 (printer port)
IRQ8	CMOS Clock
IRQ9	Redirected to IRQ2
IRQ10	Free
IRQ11	Free
IRQ12	Free
IRQ13	Math Coprocessor
IRQ14	Hard Drive Controller
IRQ15	Free

In the table 6.1 above the IRQ0 thru IRQ15 in column one correspond to the labels of connecting port physical addresses reserved for the hardware components listed under column two. As had been explained earlier any interference in any of the addresses will cause the computer to crash. There are many software applications which allow you to see which parts are using what IRQs and port addresses. One such program is MSD that comes with DOS and windows.

The Types Of Expansion Cards

Just imagine all the fancy stuff you get from desktop computers, such as video games, music, internet connections, etc. How does your computer handle them? Thanks to computer circuit cards. Nowadays the computer market is flooded with a variety of cards. But under this chapter we will discuss the most common and popular ones that are usually loaded on your standard personal computer. They are *video cards, I/O cards, controller cards, sound cards, modems, network card (NIC), memory cards, Interface Cards* and *Video Capture Card.*

Video cards (Controllers for Monitors)

Video cards also known as display adapters are circuit boards that are usually installed in motherboard to allow the monitor to display information on the computer screen. If you can see any characters printed on your computer screen, there's a better chance that your monitor is attached to a video card installed in your CPU box. Depending on the technology of the CPU architecture, the video card may slot into an ISA, PCI or AGP interface. For CPU architecture ranging from Pentium II and above it is most likely to find an AGP port. In most cases computer manufacturers will chose AGP card for high graphics quality and speed, in this case the video card will be mounted into an expansion slot popularly known as an

AGP slot (*Accelerated Graphics Port*) already discussed above. Your monitor plugs into the AGP port at the edge of the video card.

The video card allows you to see words on your computer screen. In order to display characters or images, the video card must process data received from the CPU into visual signals that may be carried to the monitor. The signals are sub-classified under ***vertical*** and ***horizontal*** frequencies. The two frequencies are fully explained under monitors later in this chapter as most monitors usually work with selected video cards. The reason being that the vertical and horizontal frequencies generated by the selected video card must fall within the range supported by the monitor. A typical vertical frequency ranges from 50 to 90Hz, while Horizontal frequency (or line rate) ranges between 31.5 KHz and 60 KHz, or more. Below are examples of video cards.

The figures 6.7 below illustrate typical Video Cards

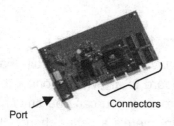

Port   Connectors

Figure 6.7a

Figure 6.7 b

To process images, video cards, like several other expansion cards, may be equipped with both mini-processor and memory chips.

Some have memory processing capacities ranging from 4-bits to 256-bits. Note the bit sequence must be multiples of 2, e.g. 4,8...

TECHNIQUES OF PROCESSING IMAGE SIGNALS

The information displayed on your monitor was processed by the video card installed in your computer motherboard which is also an electronic device. Processing activity regardless of data size requires some memory space. This explains why finding out video card memory size is important when shopping around for computers and monitors. In processing data, the video card uses the **Bit Mapping Technique.** It is bit representation of an image in the computer memory, which is mapped on to the monitor. This mapping technique is the usual way graphic adapters display information in one-colored monitor called monochrome as well as in multi-colored monitor.

Figure 6.8. Representation of bit-mapping of letter E coded in memory and corresponding image displayed on screen

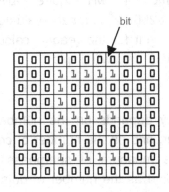

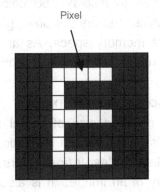

A binary representation of character E in computer RAM

The equivalent screen representation of character E

The only difference may be recorded in the number of bits that are used to carry information to one unit picture element called pixel. For monochrome monitors, only one bit of memory is used for each pixel. When the mapped bit is set to 1, the pixel is illuminated, whereas bit value set to 0 would leave the corresponding pixel non-illuminated or not visible(inactive pixel).

Displaying Colored Images

On the other hand color monitors will require more than one bit to display the same letter "**E**". Since it has multiple color capability, several bits will be used to carry all the necessary information about a pixel. The pixel will therefore need three things at this time, illumination, inactive, and color specification. For monitors that can display 16 colors, their video cards will require 4 bits of visual or graphics memory per pixel. This can be calculated as **4-bits = 2^4 = 16**. The more bits per pixel the more detailed the image will be. To be able to display 256 colors, a monitor will require 8-bits of graphics memory per pixel, given that 256= 2^8 , and 2^8 evaluates to 8-bits memory space. As an assignment for the reader, calculate how many colors will a 256-bit 4XP AGP card will display.

Resolution and Pixels

The required size of visual memory depends on the number of pixels on the screen resolution factor, and on the number of colors available. *Resolution* refers to the measure of sharpness and clarity of an image. It is also specified as the number of dots per character. *Pixel* is the smallest addressable unit on a display screen. **Pix** for picture and **El** for element = *PixEl* or pixel. Resolution is measured in unit points or pixels. For example, if you set your monitor to 640x480 resolution the quality of the image you want to display will be painted 640 points or pixels across and 480 down the screen giving a total of 307,200 pixels of bright color on your screen. If you set it to 1024x768, you will have a total of 768,432 pixels, meaning more detailed information has been added to your image, producing an output of high quality and sharp image.

The higher the pixel resolution (*the more rows and columns of pixels*), the more detailed information can be displayed. High resolution devices produce sharper, more highly defined images. Assuming that, in addition to the resolution (640x480), you set your color display to 512 colors. Then you will require at least **9 x 307,200 bits** visual memory for full screen image display. (Where $512 = 2^9$ equal to 9-bits). This works out to **345,600 bytes**, which is approximated to **350KB** memory space required on the video card. The bigger the size of the screen the more pixels will be required to display objects across and down the screen, and this implies that more memory will be required.

Sample Question. 6.1.
What is the minimum visual memory requirement for a 3-D video clip to play on monitor whose resolution is set 1024 x 768 with 65536 different colors?

Solution:

Step 1. The total number of color pixels for full screen is
```
     = 1024 x 768
     = 768,432 pixels
```
Step 2. Calculate the number of bits to hold 65536 colors per pixels
```
     65536 = 2^16 = 16-bits
     16-bits = 2bytes (where 8-bits = 1byte)
     2bytes per pixel is required
```
Step 3. Calculate total memory required for all `768,432` pixels
```
     768,432 x 2 bytes = 1,536,864 bytes.
     System will require at least 1.5MB
```

If you decide to buy a modern monitor which is bigger than 14", it is more preferable to have at least 2 Megabytes of memory on the video card. Our discussion on types of video cards under the next topic, will explain more about resolution and other measuring units for image display.

TYPES OF VIDEO CARDS - *Display Adapters*

Today's computer market is flooded with several varieties of circuitry boards that can be installed in computer motherboards to boost performance or to make the most of the fancy game packages. Video cards, often called display adapters will run in the top rank of the most preferable device when it comes to multimedia games. Like all other computer components, video cards have evolved over the years, starting from monochromes to color adapters. As of today, the most popular types have the specifications below:

- Hercules Adapter - mono color, can handle only one color
- Color Graphics Adapter (CGA) - low quality colored adapter
- Enhanced Graphics Adapter (EGA) - medium quality
- Video Graphics Array (VGA, and Super VGA) - high quality
- Extended Graphics Array (XGA) - successor of SVGA, can handle any 3-D textures of video and animation requirements of the modern multimedia programs and games. Another brand of this is the AGP or Accelerated Graphics Port.

Hercules - monochrome has been discussed above. It's only rudimentary screen display, which is not in use anymore. The only benefit for using Hercules is less memory usage. Table 6.2 below gives more elaborated information on these cards.

CGA's (Color Graphics Adapter) and EGA's – are an improvement on Hercules cards that have power to display the basic colors with less resolution. CGA could handle a maximum of 4 colors if pixel resolution is set to 320 x 200 along with text resolution of 40x25, meaning 40 characters in a row and 25 lines. When pixel resolution is set to 640 x 200 along with text at 80 x 25 only 2 colors can be displayed. The Enhanced GA , however was capable of displaying 16 colors while set to 640 x 350 along with 80 x 25 pixel and text resolutions respectively. These cards also paved the way to the creation of the VGA card series.

VGA - Stands for Video Graphics Array commonly known as video graphics adapter. The original VGA was developed by IBM in 1987 to offer clean images at higher resolutions. The standard VGA can produce as many as 256 colors at a time from a palette of 262,144 colors. To display this amount of color the original VGA, must be set to 320x400 resolution. From our previous calculations both the resolution value and number of graphic colors are limited by the amount of video memory available on the card. So at the standard 640x480 resolution, VGA was only capable of 16 colors at a time. VGA has capabilities such as "color summing" that provide some enhancements in monochrome monitors. Color summing is a process in which the VGA uses 64 different shades of gray color graphics to transform into graphics. This, in effect, simulates color on a monochrome monitor. VGA requires a VGA monitor, or one capable of accepting the analog output of a VGA card.

SVGA - Today, the VGA card is not used much, though it gave birth to more enhanced breeds. The Super VGA category of video card refers to a group of video cards, all with roughly the same capabilities(see table 6.2). It does not refer to a specific card, like the VGA technically does. SVGA was developed by third party companies in order to compete with IBM's XGA and 8514/A display adapters. SVGA is much more advanced than VGA. In most cases, one SVGA card can produce millions of colors at a choice of resolutions. But, the abilities depend on the card and the manufacturer. Since SVGA is a loose term created by several companies, there is no actual standard for SVGA.

Table 6.2. The summary of display adapters evolved over the years

Adapters	Text Resolution	Graphics Resolution	No. of Colors	Remarks
Hercules	40 x 25	Black & White	1	Monochrome
CGA	40 x 25	B & W & Color	1	
	40 x 25	320 x 200	4	
	80 x 25	640 x 200	2	
EGA	40 x 25	320 x 200	16	
	80 x 25	640 x 200	16	
	80 x 25	640 x 350	16	
VGA	80 x 25	640 x 480	2	
	80 x 25	640 x 480	16	
	40 x 25	320 x 200	256	
SVGA	80 x 25	800 x 600	256	
XGA	depends font	1024 x 768	65536	
		1280 x 1024	16-bits +	Some high-end monitors support these high resolutions. Designed for professional level work
		1600 x 1200		For computer-aided design or desktop publishing
		1800 x 1440	24-bits +	Supported by at least one: ViewSonic G810 Mega
		1920 x 1440	24-bits +	Supported by at least one: **NEC MultiSync Fe1250+** Mega

A solution meant to standardize SVGA was created by the Video Electronics Standards Association (VESA). The VESA standard for SVGA only defined a standard interface called the VESA BIOS Extension. This standard omitted certain methods of implementation of capabilities. But, it did allow programmers to write for one common interface, alleviating their task to customize individual programs to handle several different SVGA cards. It's cool! All SVGA cards in use today comply to the VESA standard. Initially the SVGA standard was distributed as programs. Enhancement made allowed them to be integrated as SVGA BIOS. Now **XGA** has tremendous capabilities to higher resolutions with millions of colors. Table 6.2 above is a summary of display adapters, and preferable resolutions and colors.

AGP Card (Accelerated Graphics Port) - It enables the graphics controller to communicate with the computer and access its main memory. The AGP is a high-speed port interface designed to handle 3-D technology, and it stores 3-D textures in the main memory rather than the video memory.

I/O Cards - These are the cards that your printer plugs into (probably your mouse too). These cards are being phased out; their components are now usually included on motherboards.

Controller Cards - This is the card that all of your drives are connected to. A lot of flat ribbons sprout off of this card. Like the I/O card, this stuff now lives on the motherboard in newer systems.

Sound Cards - These cards let your computer produce awesome sound. They are sometimes hard to install because of IRQ conflicts. In order to hear anything, you need speakers as well. Plug the speakers into the back of sound cards. The nicer the speakers, the better your sound card will sound.   An expansion board that provides audio capability and enables the computer to digitize sound beyond basic beeps.

Figure 6.9.   A Set of Speakers and a Sound Card

Sound card

Modems And Network Cards

With the modern technology of the internet and information highway advancement, we will need to spend some time to discuss modem devices. These allow you to connect to other computers via telephone lines. These things are what bring the Internet to your screen, allowing you to even see this page. Modems are considered tough to install because they need their own COM port, but I've never had much trouble with it. Also, modems have two phone jacks on the back of them. One is for the line to the wall, the other is for you to plug your phone into.

Figure 6.10. Internal Modem and an External Modem

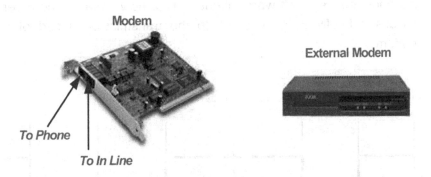

Modems are typically made to work on phone lines which transport information in wave format popularly known as analog signals. **Analog** is a form of voice, video and data transmission that employs a continual electrical signal. Modern high technology allow us to convert analog signals to digital format and vice versa.

Communication channels like telephone lines are usually analog media. Analog media is a bandwidth limited channel. In the case of telephone lines the usable bandwidth frequencies is in the range of 300 Hz to 3300 Hz. Data communication means moving digital information from one place to another through communication channels. These digital information signals have the shape of square waves and the meaning of **"0"** and **"1"**

The word **Modem** is an acronym meaning *mod*ulator and *dem*odulator, or, in other words transmitter and receiver. Modulation is the process in which digital data is converted to analog signals, making it possible for the information to travel across the phone line. Demodulation is the reverse process in which analog signals are converted into digital information which are meaningful to the human eyes. Discussion of modulation and demodulation processes alone can take a complete chapter in itself which beyond the scope this book, so we won't worry about that for now. We will however discuss it briefly to give sense to the transmission speed of a modem.

Figure 6.11. Internal Modulation and Demodulation Signals.

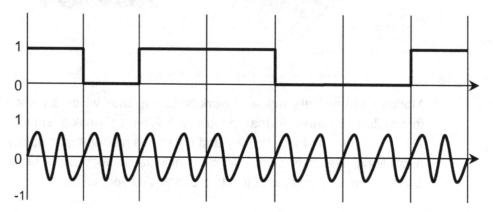

The upper diagram is a typical translation or demodulation of the analog signals below. This represents "1011001"

Modulation protocols: The specific techniques used to encode the digital bits into analog signals are called modulation protocols. The various modulation protocols define the exact methods of encoding and the data transfer speed. In fact, you cannot have a modem without modulation protocols. A modem typically supports more than one modulation protocols. These data transfer standard protocols are set by **CCITT**, a United Nations agency. It is an international telecommunications standards committee that makes recommendations on a broad range of subjects concerning data communications. The "CCITT" is a French acronym for the International Telegraph and Telephone Consultative Committee. In the United States, you will see standards like Bell 103, and 212A (300bps and 1200bps), the CCITT standards ranging from V.21-V.32, to V.90 –V.92 covering average transmission speed of **1200bps** and **56kbps** respectively.

Speed of Modem

The modem transmission speed is measured in **bit per second** or **bps.** Modems of today transmit with extremely high speed in thousands of bits per second. You will normally see **kbps** meaning **kilo-bits per second.**

3Gbps: The next generation of wireless technology that will be able to integrate voice, video and data.

28.8-kbps modem: A modem that connects to the Internet at a maximum speed of 28.8 kbps.

33.6-kbps modem: A modem that connects to the Internet at a maximum speed of 33.6 kbps.

56-kbps modem: A modem that connects to the Internet at a maximum speed of 56 kbps.

Baud

The modem transmission speed can also be measured in **baud rate.** Baud rate means **symbols per second.** There are at times a unit square wave may match several digits, for example four **"1011".** In this case if the speed is measured at 14k bauds, then the actual bits transmitted at a time will be **(4 x 14) k bps = 56 kbps.**

Network Cards Connectivity

Whether you plan to surf the Internet or send a fax, the majority of consumers purchasing a computer for their home or office want instant Internet access. This requires a modem or network interface card (NIC), which are typically standard on most systems available today. Modems usually appear in consumer-oriented PCs while NICs are found in systems designed for the home office or corporate environments. When shipped as standard features, the modem or NIC is pre-installed on the system. An important point to keep in mind is that it doesn't matter if you have the fastest CPU, your Internet experience will depend on your connection speed. Network Interface Cards (NIC) are typically digital devices.

Figure 6.12 NIC or Network Interface Card and a Switch

Switch

ADSL (Asymmetrical Digital Subscriber Line): DSL (Digital Subscriber Line) service was designed primarily for the residential consumer market. It promotes transmission of voice, video and data over copper telephone wires at very high speeds. The "asymmetric" in ADSL means that the connection transmits data at faster speeds downstream from the Internet to the computer than upstream from the computer to the Internet. ADSL can support speeds up to **8 mbps** downstream and **1 mbps** upstream, although the connection rarely approaches those figures.

Broadband: A form of data transmission in which several streams of information - data, voice and video -- can be sent at the same time over common communications lines at speeds of more than **1.5 mbps**.

key piece of telephone company technology that DSL service providers use to deliver a broadband connection to a household or business. A prospective DSL customer must be within three miles of a telephone company's central office, or switch, to receive broadband service.

Switched service: Residential or business local or long-distance telephone service that's switched, or channeled, through the local central office and the public telephone network.

Router: A device that connects computer networks to one another so that data can be ferried back and forth between and among those networks' computers. It's a piece of hardware, similar to a modem, that directs network traffic.

T-1 line: A dedicated, ultra-fast **1.54-megabits-per-second** broadband connection favored by businesses with high Internet-access and telephone-service demands.

T-3 line: A very high-speed network connection in which data is transmitted at a speed of 45 mbps.

Memory Cards: These are a way to add more memory to your computer if all of your SIMM slots are full. This memory will always work slower than the memory in the SIMM sockets because it is limited to the speed of the bus.

Interface Cards: These cards allow you to connect extra gadgets to the outside of your computer. You can get them for mice, CD-ROMs, scanners, and even adapters for laptop gadgets. If you use

MIDI for a synthesizer connected to the computer, the instrument is connected through an interface card.

Video Capture Card: You can hook a camcorder up to one of these and take pictures off the tape and save them to your computer. Also, you can buy a TV card. These allow you to watch TV on your monitor.

EXTERNAL PERIPHERALS

In the beginning of this chapter we divided computer peripheral devices into two main groups, the internal and external peripherals. We have just discussed most common internal peripherals in the preceding sub-topics as expansion cards. You now know that each internal device or expansion card provides the CPU the ability to communicate to the external world through edges known as ports. A peripheral device that must plug into such a port in order to communicate with our CPU system unit is classified here as an external peripheral. The external peripheral may therefore be identified as any computer auxiliary device that can be connected to and communicate with the computer system unit through an external port. In figure 6.1 above, you can see some common examples of such peripherals as the monitor, keyboard, mouse, printer, scanner and speakers.

By the way, these auxiliary devices operate they can be sub-classified into two distinct operational groups which are popularly known in the computer jargon as *input* and *output* devices. In fact, every computer operation can be described as being either input, output or both. Inputs and outputs are very vital topics in information technology industry. The basic function of an operating system that make the computer hardware recognize instructional sets or software, is managing input/output and interrupts requests.

Input Devices

The input device is any computer accessory that may be used to transmit information, usually characters to the computer. Typical examples are keyboard, mouse, scanner and digital camera. Keyboard and mouse are the most basic input devices for modern computers, though they may use different approach of technology in transmitting information to the memory. While the *keyboard* is used to type and write characters into the memory, the *mouse* is used purposely to capture or select information already typed and confirm transmission to the memory.

Both the keyboard and mouse control the movement of the cursor. However, while the keyboard permits cursor movements restricted in only two directions, horizontal and vertical, the mouse permits unrestricted cursor movements in all directions. This flexibility of the mouse accounts for its popularity in modern computer technology. The question now is what characters may be transmitted by an input device? We will answer this question in the next chapter.

Figure 6.13.   Modern Mice and Keyboards are input devices

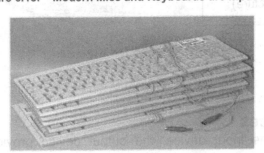

Keyboards usually come with two types of connectors, a mini 6-pin PS/2 and a larger 5-pin DIN. Keyboard adapters will allow us to convert a 6-pin PS/2 to 5-pin DIN.

Today's keyboards come with a standard number of keys ranging between 101 and 107. Some keyboards are *ergonomic;* they are designed to maintain a more natural hand and wrist posture.

Scanner – Another input device different from the keyboard and mouse. Scanners generate only mirror image of any input object it's made to process. It cannot decode the individual teletype characters in a document. It can reproduce documents and images. The scanner operates just like the printer except that it pushes data into the computer. That explains why it's a data input device.

Figure 6.14. Scanner is an input device

We put input objects here. The input object can be a document or an image (photo picture)

Output Devices

In contrast to the input device, the output device provides the means to transmit information out of the computer to its peripherals. As indicated in the computer architecture in figure 1, it must be noted that information is usually taken from the computer memory to the output devices. Computer monitors, printers, and speakers are good examples of output devices since they all carry information from the computer memory.

Monitor

The monitor or a computer screen is an output device which is capable of displaying information processed in the computer and making it eligible to the human eyes. It is the link between user and his computer. A monitor can be considered as a window into the computer's memory. It allows data entry to be checked by echoing input characters on the screen. The disadvantage of monitors is that data on the screen is temporary (called softcopy) and they can hold only limited amount of data.

Figure 6.15. The Monitor is an Output device

Modern monitors are produced using two separate architectures. A typical computer monitor may display characters by using either a single color – *monochrome* or multiples colors known as a *color monitor*. In the latter case the monitor can display a variety of colors at a time, while the former displays only one color. The color monitor uses the CGA display adapter, while the monochrome uses the MDA display adapter. (CGA stands for Color Graphic Adapter and MDA is monochrome display adapter). Depending on the number of colors that can be displayed at a time, the color monitor may be classified as VGA or SVGA (Super VGA). The fundamental display capabilities may be seen in IBM PC's and their compatibles as follows:

1. VGA – Visual graphics adapter 4-bits (2^4 = 16) colors
2. SVGA – Super visual graphic adapter 8-bits (2^8 = 256) colors
3. XGA – Extra-super visual graphic adapter 16 to 24-bits (2^{24} = 16,777216) colors. It can display over16million colors

As monitors may usually cost as much as 50% of the price of the computer, it is therefore important to know something about the hidden issues when deciding to purchase one. The quality evaluating criteria depends on the video card, the dot pitch and the maximum resolution the monitor can support.

Dot Pitch

The *dot pitch* of a monitor is the size of the pixel that is displayed on the screen. Smaller pitch values indicate sharper images and higher resolution. Most monitors have a dot pitch between 0.25 mm to 0.52mm for a good match. A dot pitch 0.28mm or smaller for 12- and 14-inch and a dot pitch of 0.31mm or smaller for 16-inch and larger monitors. Monitors of 0.39mm dot pitch usually lack clarity for fine text and graphic. It's worth noting the **dpi** (dot pitch per inch) when shopping for monitors. The smaller the *dot pitch* the higher the *dpi* value, and the *sharper* the image or the higher the resolution. It basically shows how clear the picture will be; higher dot pitch numbers may provide a fuzzy image. Anything at **0.26 mm** dot pitch or smaller should be fine for most users.

Sometimes the way the dot pitch is measured can be very confusing. It can be written as horizontal, vertical, or other forms. Usually horizontal is used, but avoid monitors that present different measurements for the dot pitch. Aperture grille monitors often have a dot pitch "range" If that is the case, look for a range of 0.25-0.27 or better (0.24-0.25).

For example we may choose a single-standard (fixed frequency) monitor and a matching video card such as VGA monitor and a VGA video card, or multisync monitor that accommodates a range of standards. Multisync monitors can accept different ranges of frequencies. So with multisync monitors, we can match the range of horizontal and vertical frequencies, which the monitor accepts with those generated by the video card. The wider the range of signals the more expensive and more versatile the monitor. The vertical (or refresh/frame rate) determines how stable the image will be. The higher the vertical frequencies the better.

Resolution

Resolution is simply the **Quality Factor** of a monitor. It defines the sharpness and clarity of an image and it can be specified in terms of the number of dots per character. We have already discussed major aspects of graphic resolutions under video cards. Currently, most software packages operate by using a lot of graphics and therefore require high resolution monitors to function properly.

It' will therefore make sense to look for monitors that support fairly high resolutions for their size. Consider, however, that it is impossible for a small 14" monitor to display at 1600x1200 resolution (same for a 15"). 17" monitors are about the minimum to be able to run at 1024x768 comfortably. A 15" monitor can usually do 1024x768, but objects are small and somewhat hard to see. 19" monitors are ideal for running at 1024x768, 1280x1024, or even 1600x1200 (although that's a little too high for my eyes).

Types of Monitors

1. Cathode Ray Tube (CRT)

Technology: A beam of electrons lights up pixels/dots on the screen, and color is achieved by combining Red/Green/Blue (RGB) of different intensities. In general CRT monitors produce brighter and better image quality.

Figure 6.16. This 17" monitor set 1024 x 768 resolution displaying a picture of 1400dpi

Size: Size is measured diagonally (corner to corner). Today monitors are available in sizes 14", 15", 17", 19", 21". But the monitor size doesn't tell anything about the maximum viewable image size. The bezel in front of every monitor's CRT diminishes the viewable area by approximately an inch.

When deciding on which monitor to buy, think of which resolution you'll be working in most frequently. The higher the resolution, the bigger the monitor is required. Monitors usually display 25 lines 80 characters each in text mode.

Standard Resolutions (in pixels)

An **interlaced** monitor draws its screen in two passes. First it draws every second line and then fills in the missing lines. Interlacing is noticeable because of flicking screen and can cause headaches.

A **non-interlaced** monitor draws its screen in one pass. Another reason for a flicking screen is the frequency with which the monitor redraws its screen - called **refresh rate** or **vertical scan rate**. The bottom line should be 75Hz so that flickering effect doesn't show up. Ensure that your monitor and graphics card can be synchronized to the same refresh rate. Higher refresh rates will definitely make the computer work easier on your eyes. Consider 75-85 Hz as a minimum refresh rate for any resolution that you actually plan to run your monitor at. Anything above 85 is a nice bonus... Make sure your **video card** will support the higher refresh rates and resolutions.

2. LCD (Liquid Crystal Display)

These displays are known as being used in calculators and watches. They have an advantage of being cheap but the major disadvantage is that it is very hard to see what they display in the dark. They are used in laptops because they are flat.

Figure 6.17.   The LCD Monitor is an Output device

A flat monitor avoids distortion of the image by the curves in the monitor. LCD Monitors are becoming more popular, but not because of better image quality. While they can be brighter, in general the image quality is better on a regular CRT (cathode ray tube) monitor. LCD monitors are really just useful because they take up less desk space. The monitor's controls can be important as well. These are helpful in tweaking the picture for proper brightness, contrast, and taking the picture to the very edges of the monitor, not to mention for removing any slight curves or other abnormalities that may be present in the picture.

3. Gas Plasma

Gas-plasma display is a type of flat display screen, called a flat-panel display, used in some portable computers. Images on gas-plasma displays generally appear as orange objects on a black background. Although gas-plasma displays produce very sharp monochrome images. They require much more power than the more common LCD displays.

Technology: A grid of conductors are sealed between two flat plates of glass; neon and/or argon gas fills the space between the plates.

PRINTERS

Printer is another output device. It makes the work of software look real. It permits reproduction of messages from the computer in the form of a printed hard copy of information processed in the computer. Like the computer monitor, the quality of characters printed could be determined by the number of printable points per inch depth also termed dpi. This measure of printing quality is known as a resolution of the printer. The higher the resolution the more accurate the character printed. Most printers communicate with computer via the parallel port. Modern technology allows printer connection through Universal Serial Bus port (USB).

BASIC TYPES OF PRINTERS

Basically, current computer technology has produced several breeds of compatible printers. We can list about three basic types:

1. Dot matrix printer
2. Ink Jet Printers
3. Laser jet printer

Dot matrix printer: An example of a dot matrix printer is the Epson LQ1000 and Plotters. The printer operates by using the printer head composed of several pins punching on printer band. You have to use a special printing paper with edge wholes that fit in a paper tractor. The paper format has continuous ends joined together to facilitate paper feed mechanism. This type of printer is counted the first generation printer more like manual typewriter. Though it's old type printer, many businesses continue to use this printer to print invoices and receipts in commercial centers.

Figure 6.18. The Dot Matrix Printer and Plotter are Output devices

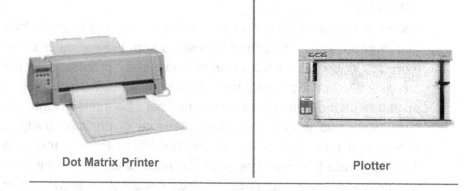

Dot Matrix Printer   Plotter

Ink Jet Printers: Example of Ink Jet Printer is HP 500 series, 600-900 Series. The inkjet printer operates by using electronic printer head spilling ink through minute holes to dynamically create printable characters. The quality of printed output depends on the size of the opened holes. Most Inkjet printers uses cartridge that is filled with ink. Some of this printer cartridges can be refilled to be used again. Unlike the dot matrix technology, inkjets mostly use standard copier paper format which sheets are not joined together.

It worth noting that freshly printed documents are often wet and, they must be left for some few seconds to dry up before touching the printed characters.

Figure 6.19. The Inkjet Printer is an Output device

Under this printer family we can identify another type know as the **bubble jet printer.** The Bubble Jets operates in the same way. These printers are mostly cheaper selling around $50-$100.

Laser jet printer: Laser jet printer is a high quality printer which operates by using powder ink cartridge. Unlike the dot matrix and inkjet printers printed documents come out of the printer already dried with high precision. It's also faster. Most business prefer Laser-jet printers for quality reasons. HP LaserJet I,II,III and IV series and all examples of LaserJet printers.

Figure 6.20.   The HP LaserJet 5100 printer is an Output device

This HP LaserJet Printer has fast printing at up to 21 ppm high-quality output with a 1200 x 1200 dpi resolution.

EXERCISE 6

1. Define a computer peripheral.

2. What's the main function of a computer peripheral device?

3. What computer component provides direct links between the system unit and the sectional peripherals?

4. How do we call the connecting ends that links the CPU to the external peripherals

5. Which of the following may not be considered as legal computer peripherals?

   A. Hard disk drive
   B. Floppy drive
   C. CPU
   D. CD-Rom Drive
   E. Expansion card

6. Expansion card provides:

   A. Link for external peripherals
   B. Is a peripheral by itself
   C. Holds the ports for external devices
   D. Not play any important role in today's computers
   E. All answers are correct except D.

7. In modern personal computers how many different types of expansion interface can be seen as vital as standard.

   A. 3   B. 4   C. 5   D. 2   E. 1

8. Which of the following cannot be classified under expansion interface.

   A. AGP   D. Memory slot

   B. PCI   E. Infra-Red[IR]

   C. ISA

9. Which tolling attributes are important to note when describing a connecting port?

   A.   Transmission mode   D. Serial or parallel

   B.   Connecting pins   E.   All of the above

   C.   Gender type

10. What is the main advantage choosing a USB port over a serial and parallel port?
    A. Flexibility by supporting multiple connecting simultaneously

    B. USB ports are normally faster
    C. USB ports requires one time port configuration by the users
    D. Handles transmission errors.
    E. No difference but just for fancy activities

11. Nowadays CPU cases have bays which are used purposely for:
    A. Loading memory chips

    B. Loading Hard drives and desired cards

    C. Loading CD-ROM/DVD drives, and floppies drives

    D. Loading Expansion ports only

    E. None of the above

12. In the answers labeled A-E which pair of peripheral devices does not match in terms of input or output
    A. Monitor/ scanner
    B. Monitor / printer
    C. Mouse / keyboard
    D. Digital cameral / joystick
    E. Mouse / scanner

13. What immediate action occurs internally in your computer system when the letter A key is pressed on your keyboard?
    A. System beeps
    B. Interrupt Request is triggered
    C. The letter A is displayed first
    D. The CPU sends signals first to your computer monitor
    E. The monitors becomes activated

14. What are the main causes for resource conflict
    A. Setting devices to a reserved IRQ
    B. Setting different devices to different IRQ
    C. Setting devices to different programs
    D. Setting more than one device to the same line of IRQ
    E. All of the above

15. IRQ lines of response is usually identified through
    A. Direct memory Access DMA
    B. Port address
    C. Caching method
    D. Swapping
    E. Virtual memory allocates

16. For a very highest resolution monitor to function well your system unit must be loaded with

    A. Interface card

    B. PCI video card

    C. Video Capture card

    D. AGP card

    E. ISA video card

17. The popular method the video card uses to process data on the screen is known as

    A. Vertical frequency sweeping

    B. Horizontal frequency sweeping

    C. Diagonal frequency sweeping

    D. Bit- mapping

    E. Gray- relayed mapping and pixel mapping

18. The measure of sharpness of an image on a monitor is determined by.

    A. Resolutions

    B. Pixels

    C. Bits

    D. Frequency horizontal / vertical

    E. All of the above

19. The clarity of color graphics is measured in:

    A. Horizontal / vertical frequencies

    B. Resolution

    C. Bits per pixel

    D. Bit per sweeping frequency

    E. Pixels only

---

20. What is the minimum visual memory required for a 3-D video clip to play on monitor whose resolution is set 1024*768 with 32-bits different colors?

    A. 0.75 MB
    B. 1.5 MB
    C. 3.25 MB
    D. 3.0 MB
    E. 4.5 MB

21. Which computer hardware component plays the role of digitizing sound from the computer?
    A. Speakers
    B. Sound cards
    C. Equalizers
    D. Sound drives
    E. Monitor

22. In order to transmute data through the line to another computer you need a modem. How does modem transmit data?

    A. Convert digital data to analog
    B. Convert analog data to digital
    C. Push analog data without converting
    D. Push digital data without conversion
    E. Demodulate data

23. Calculate the transmission speed of 28 baud rate modem in term of kbps if it was built with a capacity of 2-symbols "01.

    A. 28 kbps
    B. 14 kbps
    C. 42 kbps
    D. 56 kbps
    E. 33.3 kbps

24. Arrange in ascending order the mode of transmission commencing, with the low to high speed given the information below.
    1. ADSL   2. Broadband   3. T1   4. T3

    A.   3,1,4,2
    B.   1,2,3,4
    C.   4,3,2,1
    D.   1,3,4,2
    E.   1,2,4,3

25. Which of the following statements best describe the sharpness of image on monitor or printers?

    A.   The smaller the dot pitch the higher the dpi and lower the resolution

    B.   The smaller the dot pitch the higher the dpi and higher the resolution

    C.   The smaller the dot pitch the smaller the dpi and higher the resolution

    D.   The higher the pitch the smaller the dpi and the higher the resolution

    E.   The higher the dot pitch the higher the dpi and the higher the resolution

Software and Hardware Interface
Chapter **7**

- o **Overview Of Computer Software and hardware interface**
- o **Hardware Dependency**
- o **Role Of Bios in Software Hardware   Communications**
- o **The Hardware Dependency and Character Sets**
- o **What is Computer Program and Software?**
- o **Types of Computer Software**
- o **Bridge between Hardware and Software**

Overview Of Computer Software

In the previous chapters you have learned that the most vital operations in the computer system are inputs and outputs. We now know that the input in general sense is writing data to the memory of a computer. Modern computers are built on such architecture that before any fresh information can be stored permanently, it must initially transit through the temporary memory. In this process the memory operates as a buffer and is therefore termed as the primary memory and the permanent storage device referred to as the secondary memory.

We also learned that the **output** on the other hand provides the means to pull processed data or information out of the computer to the external peripheral devices specialized to display, print or to produce sound. Thus computer monitors, printers, speakers are good examples of output devices since they usually draw information from the computer memory.

All these fancy operations discussed above are made possible thanks to the most intelligent part of the computer system known as software. So far all the parts you have read about are visible to the human eyes. The software part is not directly visible to us, but the results of its operations can be seen as data output on appropriate hardware devices. Software is then said to be intangible, but an intellectual property. The most fundamental software required for any computer to operate is called BIOS, basic input-output system. BIOS is the foundation of all hardware dependency for all computer software. The sub-topic below will focus on the character set standards that will introduce us to modern computer programming that allow us to make most of the hardware.

BIOS (Base Input Output System)

The most fundamental software required for any computer to operate is called BIOS, basic input-output system. BIOS is composed of basic instructions needed to startup your PC. These instructions are stored in the ROM (read only memory).

Figure 7.1

AMIBIOS: Rom containing BIOS

In essence, the BIOS is responsible for booting your PC by providing a basic set of instructions. The BIOS performs all the necessary tasks that need to be initiated at start-up time: POST (Power-On Self Test, booting an operating system from Floppy disk drive - FDD or Hard Disk Drive - HDD).

Figure 7.2

Communications Between CPU and BIOS When Your PC is turned On for the first time.

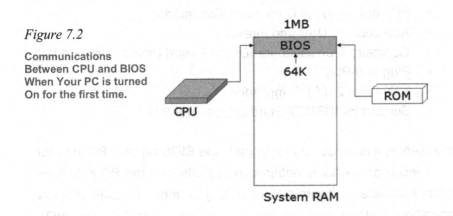

The services provided by BIOS directly affect the hardware and software (operating system). These may include, for example, disk I/O, basic character set translations and power management. However, not all operating systems use these services; some use their own instructions to access the hardware. This method of accessing the hardware may improve performance.

The Role of BIOS in Software Hardware Communications

In our definition of BIOS above we obviously mentioned some of the key role of bios, which was providing a smooth communication between both the hardware and software components. Thanks to the bios, hi-tech machines operate with some level of intelligence closer to that of human beings. The intelligence is the result of instruction set of the BIOS stored in the ROM. The instruction sets are usually written in assembler language is often called machine language will be discussed later in this chapter.

The impact of BIOS services may affect some specific component of the computer system. Some specific examples of BIOS services are:

- Activation of Standard Keyboard Character Set
- PCI or Peripheral Communication Interface
- Activation of USB and FireWire
- Geometry Translation for IDE/ATA Hard Drives Over 504 MB
- Plug-and-Play
- Y2K (Year 2000) Compliance
- Support for IDE/ATA Hard Drives Over 2 GB

In essence, it is necessary to upgrade the BIOS on your PC in order to maintain or inevitably obtain compatibility with the PC industries latest hardware, software and operating systems. In order to allow operating systems and applications to run on a PC, the **BIOS**

(Basic Input Output System) provides a standard layer of services that the operating system can use to "talk" to the hardware. In turn, the operating system provides standard services to applications to perform their functions. This means that one layer has to know how to communicate with the layer above (or below) only. This "layering" of services allows applications to run on virtually any PC regardless of the type of hardware.

Figure 7.2. The Four layers of Computer System

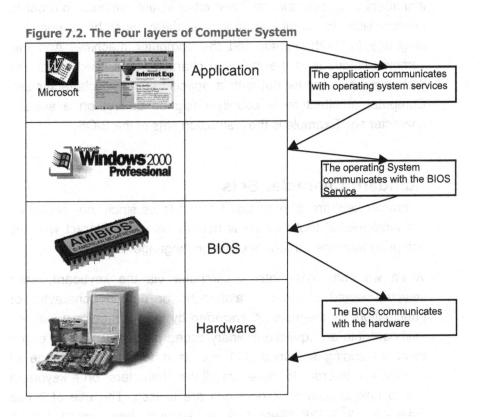

Hardware Dependency and Character Sets

Characters are images or symbols that have matching electronic signals from the computer hardware. They allow us to understand what's going on in the computer systems. In general, all operations of the computer system is character based. As was discussed in chapter three, the computer understands binary numbers (0s and 1s) corresponding to electronic signals as its characters, but humans understand character symbols like numbers (0-9), alphabets (a-z), punctuations, and other visual symbols. In order to interact with the computer machine, there must be a common language for both human and the computer machine. And this common language is the standard character sets. Each computer that is created comes out with a specific standard character set. Computer architecture is usually designed basing on a specific character set. Example is the instruction sets of the BIOS.

Standard Character Sets

Character sets are of particular interest to us since they represent the fundamental tool that allow human beings o interact with the computer machine through a common language.

When we enter data into a computer via the keyboard, each selected keyed element – alphabetic or numeric character or punctuation, for example- is *encoded* by the electronics within the keyboard into an equivalent binary-coded pattern using one of the standard coding schemes that are used for the interchange of information. In order to represent all the characters on a keyboard with a unique pattern, 7 or 8 bits are utilized. The use of 7 bits means that $2^7 = 128$ different elements can be represented, while 8 bits can represent $2^8 = 256$ elements. A similar procedure is followed on output except in this case the printer will *decode* each received binary-coded pattern to print the corresponding character, and this process is called **spooling**. The coded bit patterns to each character is referred to as **codewords.** There exist two widely used standard codes namely, **ASCII** and **EBCDIC**. The ASCII stands for American Standard Committee for Information Interchange, and EBCDIC stands for Extended Binary Coded Decimal Interchange Code.

Both coding schemes cater for all the normal *alphabetic, numeric and punctuation* characters – collectively referred to as **printable characters** - plus a range of additional *control characters* – also known as **non-printable characters**. Below are definitions for the three groups of control characters:

Table 7.1. non-printable characters

Format Control Characters		Information separators		Transmission Control Characters	
BS	Backspace	FS	file separator	SOH	start-of-heading
LF	Line feed	RS	record separator	STX	start-of-text
CR	Carriage return			ETX	end-of-text
SP	Space			ACK	Acknowledge
DEL	Delete			NAK	negative acknowledge
ESC	Escape			SYN	synchronous idle
FF	Form feed				

The EBCDIC

The EBCDIC is an 8-bit code that is used with most equipment manufactured by **IBM**. It is thus a proprietary code but, owing to the widespread IBM equipment in the computer industry, it is frequently used. Figure 7.2. below is a table representing the standard interchange codes known as **EBCDIC**. The characters are represented by 8-bits code words allowing **256** character symbols. In standard interchange codes, all character sets are represented by seven bits. In EBCDIC, the remaining 1-bit at position 8 serves two purposes. The first one is representing the parity bit which is used for the purpose of error-detecting in data transmission. The second use is being part of the numerical value of the character code. That's when bit-8 =1, then its actual numeric value will be 2^{8-1} = 2^7 = 128_{10} .

Figure 7.2. Standard Interchange codes: **EBCDIC**

The EBCDIC Table

Bit Position — the four bit rows (4, 3, 2, 1) give the bit pattern of each data column:

Bit Position	col	col	col	col	col	col	col	col	col	col	col	col	col	col	col	col
4	0	0	0	0	0	0	0	0	1	1	1	1	1	1	1	1
3	0	0	0	0	1	1	1	1	0	0	0	0	1	1	1	1
2	0	0	1	1	0	0	1	1	0	0	1	1	0	0	1	1
1	0	1	0	1	0	1	0	1	0	1	0	1	0	1	0	1

Main table (row label = bit positions 8 7 6 5):

8765	0000	0001	0010	0011	0100	0101	0110	0111	1000	1001	1010	1011	1100	1101	1110	1111
0000	NUL	SOH	STX	ETX	PF	HT	LC	DEL			SMM	VT	FF	CR	SO	SI
0001	DLE	DC1	DC2	DC3	RES	NL	BS	IL	CAN	EM	CC		IFS	IGS	IRS	IUS
0010	DS	SOS	FS		BYP	LF	EOB	PRE			SM			ENQ	ACK	BEL
0011			SYN		PN	RS	UC	EOT					DC4	NAK		SUB
0100	SP										¢	.	<	(	+	`|`
0101	&										!	$	*	)	;	¬
0110	-	/									'		%	-	>	?
0111											:	#	@	,	=	"
1000		a	b	c	d	e	f	g	h	i						
1001		j	k	l	m	n	o	p	q	r						
1010			s	t	u	v	w	x	y	z						
1011																
1100		A	B	C	D	E	F	G	H	I						
1101		J	K	L	M	N	O	P	Q	R						
1110			S	T	U	V	W	X	Y	Z						
1111	0	1	2	3	4	5	6	7	8	9						

As you can see from the table above, the 8-bit positions are divided into two groups upper and lower positions as indicated at the top left corner. The lower bit positions starts from 1 to 4 upwards in vertical direction. The bit values in this group can be read from each position horizontally across the table along the line indicating the position number. The upper bit positions on the other hand starts from 5 to 8 in horizontal direction counting from left to right. The bit values in this group can be read from each position vertically down the table along the line indicating the position number.

How to read the EBCDIC table

Supposing we want to read the character "**A**", from the EBCDIC table in figure 7.2. We may first find A in the table. Then we write the binary digits corresponding to the Bit Positions starting from 1 through 8 as follows:

> **It must be noted that the calculation and conversion to decimal base numeral is based on the bit position. The bit position of the binary digit is the rank counting from left to right. For example the bit position 1 corresponds to the binary digit 1 and thus the value is**
>
> $2^{1-1} \times 1 = 2^0 \times 1 = 1.$ *Bit position 7 is equal to*
> $2^{7-1} \times 1 = 2^6 \times 1 = 64,$ *and the bit position 5 is equal to*
> $2^{5-1} \times 0 = 2^4 \times 0 = 0$

Bit positions 8 7 6 5 4 3 2 1

A = 1 1 0 0 0 0 0 1 => $1\ 1\ 0\ 0\ 0\ 0\ 0\ 1_2$

7 6 5 4 3 2 1 0

A = | 1 | 1 0 0 0 0 0 1 $1\ 1\ 0\ 0\ 0\ 0\ 0\ 1_2$

parity bit 128_{10} + 65_{10} => $(2^7 \times 1) + (2^6 \times 1) + (2^0 \times 1)$

(error-detecting bit)   => 193_{10}

The character "**A**" can be represented by the codeword **11000001** and having a corresponding numeric value of **193** (*128+65*). This means that each time the user press on the key corresponding to the character "**A**" is the number **193** is transmitted to the keyboard buffer, then the PCU process it as 11000001 and store in the memory and transmit the symbol "**A**" to the output port to be displayed on the monitor. The character "**a**" can also be represented by the codeword **10000001**, evaluating to a numeric value of **129**. Refer to the appendix in this book for the exact comparison table for **ASCII** and **EBCDIC**.

Parity bit

The reader might notice that the bit-8 or the digit at position 8 was not use in any of the calculations. This bit is used for the detection of errors in transmission. It is therefore known as the parity bit. The reader may refer to any data transmission book for more information of parity bit.

The ASCII

The ASCII code is the same as that defined by the ITU-T (International Telecommunications Union – Telecommunications (sector)), known as **International Alphabet Number 5 (IA5).** And also that used by the International Standards Organization known as **ISO 645.** Each character is a **7-bit** code, therefore allowing a representation of only 128 characters. Character sets transmitted by an input or output device are defined under standard codes defined above. The ASCII code not only facilitates internal communications among the components of the computer, but also makes it possible for different computers to communicate.

Similarly, figure 7.3 is an example of a table that represents the standard interchange codes known as *ASCII*. In contrast with the

EBCDIC the characters are represented by 7-bits code-words. It is a popular character representation, which does not include error-detecting bit.

Figure 7.3. Standard Interchange codes: (b) **ASCII**

The ASCII Table

Bit positions			7	0	0	0	0	1	1	1	1
			6	0	0	1	1	0	0	1	1
			5	0	1	0	1	0	1	0	1
4	3	2	1								
0	0	0	0	NUL	DLE	SP	0	@	P	\	p
0	0	0	1	SOH	DC1	!	1	A	Q	a	q
0	0	1	0	STX	DC2	"	2	B	R	b	r
0	0	1	1	ETX	DC3	#	3	C	S	c	s
0	1	0	0	EOT	DC4	$	4	D	T	d	t
0	1	0	1	ENQ	NAK	%	5	E	U	e	u
0	1	1	0	ACK	SYN	&	6	F	V	f	v
0	1	1	1	BEL	ETB	'	7	G	W	g	w
1	0	0	0	BS	CAN	(	8	H	X	h	x
1	0	0	1	HT	EM	)	9	I	Y	i	y
1	0	1	0	LF	SUB	*	:	J	Z	j	z
1	0	1	1	VT	ESC	+	;	K	[	k	{
1	1	0	0	FF	FS	,	<	L	\	l		
1	1	0	1	CR	GS	-	=	M	]	m	}
1	1	1	0	SO	RS	.	>	N	^	n	~
1	1	1	1	SI	US	/	?	O	_	o	DEL

Like the EBCDIC table above, the 7-bit positions of the ASCII table are divided into two groups upper and lower positions as indicated at the top left corner. The lower bit positions starts from 1 to 4 in horizontal direction counting from left to right. The bit values in this group can be read from each position vertically down the table along the line indicating the position number.

The upper bit positions on the other hand starts from 5 to 7 inclusive upwards in a vertical direction. The bit values in this group can be read from each position horizontally across the table along the line indicating the position number.

How to Read the ASCII Table

The characters in this table can be read in the same way as have been described above. Taking the example of the character **"A"** from the table of figure 7.3, we can first find the symbol A in the table. We then write the binary digits corresponding to the Bit positions in both vertical and horizontal directions using "A" as the point of reference. We can therefore select the bits starting from positions 1 through 7 as follows:

Figure 7.4. Illustrates how to read the character "A" from the ASCII table

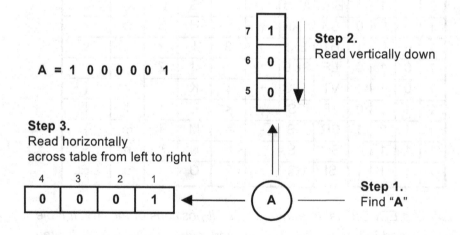

Now follow the steps below to evaluate the binary numbers into standard decimal numbers (0-9).

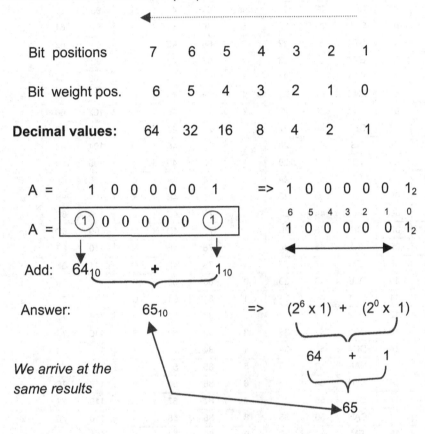

As we have already seen the character "**A**" can be represented by the codeword **1000001** which works out to a corresponding ASCII value of **65**. The reader must note that in ASCII codeword there is no digit at position 8 and thus there is no error detection and no parity bit. The figure 7.5 below is another form of ASCII table list corresponding values in both decimals and hexadecimals.

Figure 7.5. ASCII Decimal And Hexadecimal Table

Dec	Hex	Code	Dec	Hex	Code	Dec	Hex	Code	Dec	Hex	Code
0	00	NUL	32	20	space	64	40	@	96	60	`
1	01	SOH	33	21	!	65	41	A	97	61	a
2	02	STX	34	22	"	66	42	B	98	62	b
3	03	ETX	35	23	#	67	43	C	99	63	c
4	04	EOT	36	24	&	68	44	D	100	64	d
5	05	ENQ	37	25	%	69	45	E	101	65	e
6	06	ACK	38	26	$	70	46	F	102	66	f
7	07	BEL	39	27	'	71	47	G	103	67	g
8	08	BS	40	28	(	72	48	H	104	68	h
9	09	HT	41	29	)	73	49	I	105	69	i
10	0A	LF	42	2A	*	74	4A	J	106	6A	j
11	0B	VT	43	2B	+	75	4B	K	107	6B	k
12	0C	FF	44	2C	,	76	4C	L	108	6C	l
13	0D	CR	45	2D	-	77	4D	M	109	6D	m
14	0E	SO	46	2E	.	78	4E	N	110	6E	n
15	0F	SI	47	2F	/	79	4F	O	111	6F	o
16	10	DLE	48	30	0	80	50	P	112	70	p
17	11	DC1	49	31	1	81	51	Q	113	71	q
18	12	DC2	50	32	2	82	52	R	114	72	r
19	13	DC3	51	33	3	83	53	S	115	73	s
20	14	DC4	52	34	4	84	54	T	116	74	t
21	15	NAK	53	35	5	85	55	U	117	75	u
22	16	SYN	54	36	6	86	56	V	118	76	v
23	17	ETB	55	37	7	87	57	W	119	77	w
24	18	CAN	56	38	8	88	58	X	140	78	x
25	19	EM	57	39	9	89	59	Y	121	79	y
26	1A	SUB	58	3A	:	90	5A	Z	122	7A	z
27	1B	ESC	59	3B	;	91	5B	[	123	7B	{
28	1C	FS	60	3C	<	92	5C	\	124	7C		
29	1D	GS	61	3D	=	93	5D	]	125	7D	}
30	1E	RS	62	3E	>	94	5E	^	126	7E	~
31	1F	US	63	3F	?	95	5F	_	127	7F	DEL

Unicode- Storing Data

The list of character codes mentioned above is not exhaustive. There are several standard codes in common use for representing characters. Some computer software, however store characters in pattern of 16-bits and 32-bits code-words. The most popular codeword is the 16-bits codeword commonly known as **Unicode**. Unicode is International Standards Organization (ISO) character standard. Unicode uses a 16-bit (2-byte) coding scheme that allows for 65,536 distinct character spaces. Unicode includes representations for punctuation marks, mathematical symbols, and dingbats, with substantial room for future expansion. In the Unicode format a character requires two bytes of storage space. Unicode is mostly used in Java programming language. For example, in Unicode format   the name **"Kofi Siaw"** would be represented by the following values:

K	0000000001001011
o	0000000001101111
f	0000000001100110
i	0000000001101001
Space	0000000000100000
S	0000000001010011
i	0000000001101001
a	0000000001100001
w	0000000001110111

Note that the space is considered to be a character. A single bit doesn't contain much information, so most computers group bits into larger units called bytes, which usually contain eight bits. A number typically occupies four bytes (32 bits). A Unicode character occupies two bytes (16 bits). All software works on the basis of computer standard character set. Prior discussion on character set will certainly set the tone to the subsequent topic on software. The bells and whistles of the present high technologies have much to thank the computer software.

Software

In chapter two above you saw in figure 2.1, an illustration of a standard personal computer with accessories. It appears therefore to be all that we know about a personal computer. The designers follow a whole bunch of standard abstraction approach that makes a computer looks like a simple machine. There are a whole lot of hidden complex components we've got the chance to discover throughout the preceding six chapters. In other words it is only the hardware or the solid physical part that users can see. The invisible part that makes that machine functions as it is supposed to be, is known as the software.

What is Computer Software ?

Software consists of programs that instruct the hardware how to perform operations. A program is simply a step-by-step set of instructions that tells a computer what to do. Actually, a software is a collection of programs working together to resolve a predefined problem. For the above reasons a thorough definition of a program will provide the true meaning of a software. This means, that a software is simply a computer program. It will therefore make sense to take a moment to briefly discuss a program.

If you have used a computer before, you have used a program. As I have already mentioned in chapter one, a computer system is a combination of a software and hardware. For a computer to work both parts must be operational. The software part represents a set of programs. Perhaps you have written a letter using a program like the Microsoft Word, played a computer game, or balanced your checkbook with Microsoft Excel or with Quicken. All of these are programs, which were created by writing lines of commands, called *codes*. Each line of code contains instructions that dictate what should happen given a certain kind of input. For example, somewhere in Microsoft Word's code there is a line that says "if the user chooses Arial font with 12 points, change the text so that it is displayed in Arial 12 points".

Types of Software

As the development of software may be very tedious and time consuming, designers usually set scope limits for each software to address a specific problem. The purpose for which the software was developed usually defines which category it falls into. There are software that prepare and make certain hardware components usable by other software, while others activate such hardware devices as the monitor and keyboard allowing users the possibility to interact directly with the computer system. Most software falls into one of the three categories below:

- Operating System. Example, DOS 6.0 and Windows 98, 2000, XP, Unix, and Linux
- System Software. Example, Compiler and interpreter
- Application Software. Example, Microsoft Word

Operating System Software

Each system you look at should come with some software, including an operating system and various applications. The operating system is the software that coordinates communications among various system components. The operating system is a rather large and elaborate collection of programs that interact directly with the computer's hardware. It's the next level software directly above the BIOS. At present, common operating systems include DOS (disk operating system), windows, MacOS (Macintosh Operating System) ,Unix and Linux also a version of Unix, which is becoming increasingly popular. Most personal computer (PC) systems will offer one of two operating systems from Microsoft: Windows NT,2000 or Windows 98,Me,and XP. If you are using the PC in a professional environment, Windows NT or 2000 is the operating system to choose because it has a built-in security and management features that are important to a business. Windows 98, Me or XP, on the other hand, are all examples of operating systems designed for typical consumers. These operating systems offer support for thousands of hardware peripherals and boast a similar interface. The services provided by the Operating systems are so crucial that, no reasonable data processing can be carried out without them. This explains why the founder of the Microsoft Corporation, Mr. Bill Gates stands out to be the wealthiest person in the whole world. Imagine every single piece of personal computer manufactured in this world will require at least one of Mr. Gates operating systems in order to function.

System Software

The system software is closer to the operating system. It's any software other than the operating system that can be used to create another software or directly be used to manage hardware components. It's worth noting that the system software will not operate on any computer without an operating system installed in it. Examples of system software are compilers, and drivers. A driver is any software package that will activate a piece of hardware component so it can interact with the operating system. Every hardware component requires a driver software to operate properly. This sub-topic will focus on compilers or programming languages.

Compilers and programming languages

One of the most amazing aspect of computing is the ability to design your own programs. As was defined earlier, a program is a set of orderly instructions that tells a computer what to do. Unfortunately, programs are not written in any human spoken language such as English. They are rather written in languages that are easy for computers to understand. And since the computers are the ones that have to do most of the work, I suppose that makes sense. Some common programming languages include Assembler, C, C++, Pascal, Ada, Java, Visual Basic, Java script, and Perl. Though any of the programming languages uses English characters, the ultimate program is translated in binary numbers for the computer to understand. This machine language is usually generated by a system software called **compiler**. The reader may refer to my second book entitled, "Introduction to Object Oriented Programming – using Java", for more in depth discussion on compilers and interpreters.

These programming languages may be classified to generations depending on how closer the code to human language (English). The diagram below illustrates the five common levels for computer language evolution.

Figure 7.6. Generations of Programming Languages

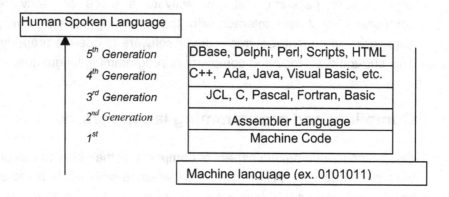

Every computer comes with such a language, known as machine language. Which machine language your computer supports depends on the type of CPU installed in it. Intel CPUs understand one language, Motorola CPUs use another, and so on. The language used by different CPUs from the same manufacturer is often similar, although not exactly the same. For example, programs that will run Intel's 486 CPU will as well run on a Pentium CPU, but not all Pentium programs will run on a 486. Although it's possible to write programs in machine language, that's not what programmers typically do. Machine language is extremely primitive, making it difficult to write even simple programs.

Instead, most programmers use high-level languages that aren't tied to a particular computer. There are a number of high-level languages in common use; here are a few you may have heard of:

- **Ada** (named for Ada Lovelace, a 19th-century protégée of Charles Babbage, who invented a mechanical computer). Developed in the late 1970s for the U.S. Department of Defense, Ada is now widely used in the defense and aerospace industries.
- **BASIC -** (Beginner's All-purpose Symbolic Instruction Code). Developed in the 1960s for programming novices, BASIC is now widely used for developing commercial applications. Many dialects of BASIC exist, of which Microsoft's Visual Basic is the most popular one.

- **C -** Developed by Bell Laboratories in the 1970s in conjunction with the Unix operating system, C is often used for programs that need to be very fast or run in a limited amount of memory.

- **C++ -** Developed during the 1980s at Bell Laboratories, C++ is a more modern, "object-oriented" version of C.

- **COBOL -** (Common Business-Oriented Language.). developed in the early 1960s for business applications, COBOL has long been a mainstay of the business world.

- **FORTRAN -** (*For*mula *TRAN*slation). Developed by IBM in the 1950s, FortRAN was the first widely successful high-level programming language.

- **Pascal -** (named after Blaise Pascal, an 18th-century mathematician and inventor of the first calculator). Designed by Swiss computer scientist Niklaus With in the early 1970s, Pascal became the leading language for teaching computer science during the 1980s.

- **Java -** is a high-level language that was introduced in 1995. Java was created by a Canadian called James Gosling while working in Sun lab. Java has now become one of the best-known programming languages on the planet. Java is famous for allowing Web developers to write cool " applets"—small programs embedded in web pages. But Java's significance goes beyond that, as we'll see later. My second book is on modern Object Oriented Programming using Java.

Application Software

An application is a program designed to perform useful tasks at a high level users interface. Most application software are not intended for programming purpose, however some of them come with macro programmable interface. Macro consists of a few lines of program instruction sets. For example, a word processing program is an application, and so is a program that allows a person to draw pictures, send electronic mail, or play a game. Applications include software development tools---programs. An operating system serves as a bridge between hardware and applications. It provides services that most applications will need: obtaining input from the keyboard and mouse, displaying text and graphics on the screen, storing data in files, sending data to the printer, and so on. Applications are therefore much easier to write because they don't need to deal with such low-level details. When an application needs to interact with the hardware, it sends a request to the operating system which performs the operation on behalf of the core of the application. We can visualize the hardware as the core of a computer system. The operating system is a layer that surrounds the core, and the applications form another layer around the operating system. The figure below is a simplified version of our illustration of the four layers in figure 7.2.

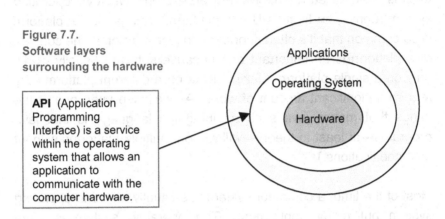

Figure 7.7.
Software layers
surrounding the hardware

API (Application Programming Interface) is a service within the operating system that allows an application to communicate with the computer hardware.

As shown in the diagram above the application components interact with the hardware through application programming interface (API) tools built into the operating systems. In advanced Visual Basic programming API's may be in the form of dynamic link library (DLL) that can be called from the operating system. Depending on which operating system is running behind the scenes, the API will be classified under 16-bit or 32-bit. API's embedded in Windows NT 4 and later versions are usually 32-bits, whereas those embedded in the previous versions are known as 16-bit API's. Similarly, the programming languages like Visual Basic 6.0 or **.Net** will only run 32-bit API's while the previous versions are only compatible with 16-bit API's.

Platform

The combination of an operating system and a particular type of CPU is often called a platform. For an example windows operating system running an Intel CPU is a platform. (This particular platform is so common that it's often shortened to just "**Wintel**"). The concept of a platform is an important one because software usually works only on a single platform. Making it run on a different platform may require a significant amount of work. (As we'll see later, one of the things that makes java such an intriguing language is that java programs---at least in theory—will run on multiple platforms without any modifications.)

Most of the time, a computer system has usually only one operating system but many applications. The operating system is often preinstalled when a computer system is purchased, and it's not often changed, although it may be upgraded occasionally. Applications on the other hand usually installed by the user can be removed and reinstall new ones at any time. An operating system often has a number of different versions, which can differ considerably. Common versions of Windows, for example, include windows 3.1-3.5, Windows NT, Windows ME, 2000 and XP, Windows CE, and whatever new versions Microsoft has released since this book was published. Unix and Linux are also operating systems. Applications are usually designed for one particular version of the operating system. Applications written for an older version of an operating system will often—but not always--work properly with a newer version.

Manufacturers also are spicing up systems by including versatile application suites, such as Microsoft Office or Microsoft Works. Some systems ship with a dozen or more extra programs, including everything from business packages to games. Buying a system strictly because it has a colossal software package isn't always a good idea, but when comparing similar systems at similar prices, a great software bundle may influence your decision.

Ways of Interacting with Computers

Most applications need to communicate, or "interface," with the user by displaying information for the user to see and accepting commands from the user. There are two primary types of user interfaces: graphical user interfaces and text-based interfaces. Let's take a look at the differences, between these two types of interfaces. I'll use Windows as an example, but the principles are the same for other platforms.

Graphical User Interfaces

Most applications now rely on a **graphical user interface**, or GUI (pronounced "gooey") built out of visual components. When such a program is running, it displays a Window on the screen. For example, if we run Microsoft word – a word processor— we'll see a window similar to the one below. This window is composed of thousands of tiny pixels (picture elements), each with its own color. (The exact appearance of the window depends on the version of the word application; your version may display a window that's somewhat different.)

Figure 7.8. An Example Microsoft Word window.

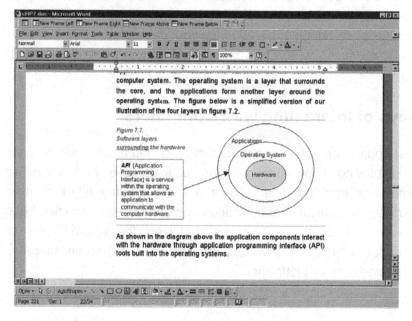

Within a GUI window are visual Components that display information to the user and (in most cases) allow the user to perform an action. In the word window, we see the following features:
- Button
- Menu
- Popup
- Text area

Buttons. This program has two rows of buttons. Each represents a particular command, which can be selected by pointing at the button using the mouse and then clicking a mouse button. Usually the most common commands are assigned buttons.

- **Menu.** Most programs have so many commands that there isn't enough room for each one to have a button. Instead, the user must click on one of the words listed at the top of the window. Doing so will cause a menu of commands to appear. The user can then choose one of the commands from the menu by clicking a mouse button.

- **Popup menu.** In some cases, the user is allowed to make a choice from among a list of possibilities. One of Word's popup menus allows the user to choose a particular typeface (font). The user presses the tiny arrow button at the right side of the menu to view the choices and then selects one with the mouse.

- **Text areas.** The empty space in the middle of the window is a text area. The user can display a document here and make changes to it. Performing certain action will cause other windows to appear. These windows, known as dialog boxes or dialogs, are used to display information to the user and/or accept input from the user. For example, if we select the Open command from the file menu or simply click on the second button from the left, we'll see the dialog box shown at the top of the next page. This particular type of dialog box is called a file dialog box or a file dialog. A file dialog allows the user to choose a file name, either by typing it or by "browsing" the computer's disk drives and then making a selection.

- **Minimizing.** Clicking on the –button makes a window disappear temporarily from the screen. In Windows, an icon representing the window appears in the taskbar, which is normally located at the bottom of the screen. By clicking on this icon, the user can later restore the window.

- **Maximizing.** Clicking on the button will cause a window to expand to its maximum size.

- **Restoring.** Clicking on the button will cause a window to return to its original size (before it was maximized).

- **Closing.** Clicking on the x button closes a window. For example, restoring the Microsoft Word window shown earlier has the following result on my computer:

Notice that the window can't be reduced further without losing space for some of the buttons at the top.

Graphical user interfaces—as we know them—were developed during the 1970s at Xerox Corporation's Palo Alto Research Center (PARC). Researchers there developed the first workstation computer, the Xerox Alto. Unlike most computers of that era, the Alto was designed for a single person to use. The Alto had a high-resolution display that could show images as well as text. To allow the user to work with more than one program at a time, Xerox researchers came up with the idea of displaying multiple windows on the screen. Windows not currently displayed at their full size were represented by icons. To make it easier for the user to interact with the interface, a mouse was added. The result was quite similar to the graphical user interfaces of today.

Graphical user interfaces weren't an immediate success. Xerox's attempt to sell computers with graphical user interfaces was a failure. Apple computer, headed by Steve Jobs at the time, saw the advantages of the GUI approach and incorporated it into a computer called the Lisa, which also flopped. Fortunately, Apple tried again and made a success of the Macintosh, the first widely used computer to support a graphical user interface. Users of the IBM

---

personal computer and its clones had to wait a while longer, until Microsoft Corporation developed Windows, a graphical user interface for DOS users. Even then, many DOS users derided the GUI approach, referring to it as a WIMP (Windows, Icons, Mouse, Point) interface. Today, of course, graphical user interfaces are dominant, to the extent that many users are able to avoid text-based interfaces entirely.

Text-Based Interfaces

Before the advent of graphical user interfaces, programs used a text-based interface, in which all input and output consisted of characters. In a text-based interface, no graphics are displayed, and user commands are entered from the keyboard. Programs that are written for **DOS** (Disk Operating System) usually rely on a text-based interface. Text-based programs are normally run from a Command line. The operating system "prompts" the user to enter a command, which is then executed by the operating system. A command could represent a "built-in" operating system capability, (listing the files in a directory, for example), or it could be a request to execute a program installed (or written by) the user. If the user enters the name of a program, the operating system will " load" the program and start it running. The form of the prompt varies, depending on the operating system. In Unix and its variants, such as Linux, the prompt is often—but not always—a percent sign:

In DOS, the prompt is a **">"** character, preceded by a letter indicating which disk drive is the " current" one: **C:\>**
The DOS prompt is often configured to display the " current directory" as well: C:/windows>

The operating system that provides a graphical user interface may still allow the user to run text-based programs from a command line. In Windows, for example, the user can open a "DOS window," which allows programs to be run from a DOS-like command line. This can be achieve by selecting **Run** from the start menu in windows and typing "**cmd**".

A DOS window is normally capable of displaying up to 25 lines, with each line limited to 80 characters. Several DOS windows can be open at a time in windows environment, which can be quite convenient.

We can run a text-based program within a DOS window by simply typing the name of the program. For example, DOS provides a program named edit that can edit files. To run the program, we would type edit and press the **Enter key**. The DOS window would now have the appearance shown at top of the next page. The edit program may look "graphical," but it's still considered to be a text-based application. The lines that you see are actually formed out of special characters. The reader can refer to any of Microsoft Operating system user manual for information.

EXERCISE 7.
Answer all questions.

1. What is BIOS ?
   a. Describe the role of BIOS in a computer system
   b. What is the difference between BIOS and an Operating System?

2. Distinguish between a computer program and a platform.

3. What is an Algorithm in computing?
   a. Describe the importance of Algorithm in software designing
   b. Give an example a simple algorithm

4. In your own words define a pseudo-code

5. Distinguish between an algorithm and pseudo-code.

6. What does a GUI stand for?
   a. Give some examples on GUI in modern computing.
   b. What operating systems most support GUI designing in software development.

7. What is a programming language?

8. Compare and contrast Operating system, Compiler, and Application Software.

9. What is a variable in computing?

10. Describe the role variables play in computer programming.

Software - Basic Elements Of Programming

Chapter 8

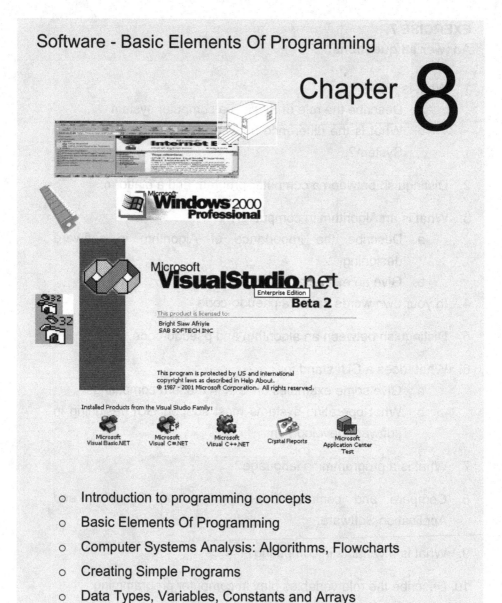

o Introduction to programming concepts

o Basic Elements Of Programming

o Computer Systems Analysis: Algorithms, Flowcharts

o Creating Simple Programs

o Data Types, Variables, Constants and Arrays

o Loops and Conditional decisions in programs

Overview of Basic elements of Programming

This section will focus on discussions pertaining to the basic principles readers would need to know about computer programming. It will outline the key elements on programming methodologies providing some highlights on analysis, designing, coding and testing procedures.

What is Computer Programming?

In a nutshell, programming means writing down a series of instructions that tell a computer what you want it to do. These instructions have the following properties:

- *Computation proceeds in discrete steps.*
- *Each step is precisely defined to provide sufficient computing details.*
- *The order in which steps are performed may be important.*
  A set of instructions with these properties is said to be an algorithm. The steps in an algorithm can be short and simple, but they need not be redundant.
- *Some steps may involve a series of smaller steps.*
- *Some steps may involve making decisions.*
- *Some steps may repeat.*

Algorithms are common in the real world. A cooking recipe is an algorithm, and so it is for a set of instructions that explains how to build a bookcase or read your email on the internet.

Let's attempt to prepare one of my favorite dishes popularly know as "**Fufu**" and palm-nut soup. Fufu meal has two parts, the soup or gravy and pounded yam or mashed potatoes. It's usually preferable to begin with the preparation of soup.

Preparing Palm-Nut Soup:

1. Clean 2 lbs of meat
2. Chop the meat into 10 pieces
3. Chop 1 onion and 2 fresh tomatoes
4. Add:
    a. chopped meat and vegetable in a deep saucepan
    b. ½ tablespoonful of seasoning spices
    c. 1½ tablespoonful of salt.
5. Steam at 350 F for 10 minutes
6. Add 40 oz canned palm nut juice.
7. Allow it to boil for 25 minutes while stirring in every 5 minutes.
8. Now add ¾ liter of water
9. Allow to cook for 25 minutes

Preparing Fufu:

1. Slightly warm ¼ litre of water
2. Soak 2 tablespoons of potato starch in warm water
3. Stir thoroughly to obtain a homogenous mixture
4. Soak ½ lb of mashed potatoes in ½ liter of cold water
5. Add products of steps 3 and 4 and stir to obtain a mixture
6. In a plastic container cook in microwave for 10 minutes
7. Or cook in a cooking port at 350 F and stir every 1½ minutes for 10-15 minutes.
8. Stir and beat to make it more palatable
9. Mould into serving sizes with soup.
10. Serve at once

This recipe satisfies most of the requirements for an algorithm:

- Algorithm involves discrete steps.
- It also involves a decision making, and how often one is required to stir to obtain a homogenous mixture.

Although this recipe is detailed enough for an experienced cook, it might pose some problems for a novice. A computer is the ultimate novice—it does exactly what you tell it to do, even if you tell it to do something that's blatantly wrong. For that reason, our "*recipes*" (*algorithms*) will need to be much more precise than a Joy of cooking recipe.

Computer Algorithms

As was discussed in chapter one, computer like any other machine follow the **I-P-O** process. This means computer follows the famous input – process – output method which clearly outlines the principle of algorithm.

Computer algorithms often involve obtaining data inputs, performing calculations, and producing outputs. Let's consider a simple problem that can be solved by a computer: calculating the biweekly salary of an employee. Here are some of the steps that best represent the salary calculating process:

1. Display a message asking the user to enter the timesheet information and wage rates.
2. Obtain the input entered by the user.
3. Convert the user's inputs into appropriate numerical forms.
4. Calculate the biweekly salary, using the following formula **S = (Hours) x (Rate)**
5. Convert salary calculated into character form and display the result.

I have divided the algorithm into discrete steps. Most of these steps are reasonably precise, although it's not clear exactly at this stage how we are going to display the information to the user and obtain the user's input. That will depend on what type of program we eventually write. In a GUI application, input usually takes the form of clicking a button, making a choice from a menu, or typing characters into a box. In a text-based application, input comes from the keyboard. Another issue that's a bit fuzzy is **step 3:** converting the user's input to numerical form. What action do we take if the input is not in the form of a number? It would be nice to assume that users never make mistakes, but that's rarely the case in the real world. We'll have to decide whether the algorithm should stop at that point, or whether it should inform the user of the problem and ask for new input.

Expressing Algorithms

There are a number of ways to express algorithms; we will use three of them in this book:

- **Natural languages**. Algorithms can be written in a natural (human) language—recipes in cookbooks are expressed in this way. The advantage of natural language is that anyone who understands the language can read the algorithm. However, natural languages often lack the precision that we will need for expressing algorithms. Also, computers have troubles understanding natural languages.

- **Programming languages**. In order for a computer to be able to execute our algorithm, we will need to express it in a programming language. Programming languages provide the necessary precisions that are simple enough for computers to understand and interpret. (Whether they're simple enough for humans to understand is another matter.)

- **Pseudo-code**. Pseudo-code is a mixture of natural language and a programming language. An algorithm written in pseudo-code is more precise than one written in natural language but less precise than one written in a programming language. On the other hand, a pseudo-code algorithm is often easier to read (and to write) than one expressed in a programming language. Usually as best practice a clear programming details will often be expressed in a form of a pseudo-code, before translating into an actual programming language. In this way algorithms and pseudo-codes guide the programmers to avoid serious semantic errors that can be overlooked during coding. Semantic errors occur mostly during runtime when all syntax errors are correct but the logic of processing is incorrect, resulting in an unexpected output.

Variables

Variables are memory locations where data are stored during the execution of a program. Our payroll algorithm for an example will have to store four items of data in runtime:

- The input entered by a user
- The hours worked
- The wage per hours
- The Salary value

These are the locations that are used to store data within a program, and they are known as variables. Variables are given names by the programmer. We can choose whatever names we want, subject to the rules of the programming language. It is best to choose a name that suggests what data the variable stores. For example, we might store the Salary amount in a variable named **BiweeklySalary**,

SalaryTemp, or just Salary. Shorter names are usually less descriptive and therefore undesirable. "*Sal*" would be worse than *Salary*, and *S* would be even worst still. Using a name that's completely unrelated to the value that it represents (such as *a* or *x*) would be in unspeakably poor taste; names such as these provide no useful information to anyone who might read the program in the future. When choosing a name for a variable, pick one that suggests what data the variable stores. Avoid names that are too short, and unrelated to the values intended for storage in the variables, or have more than one obvious interpretation.

Another example of a simple algorithm is the **Fahrenheit-to-Celsius** algorithm illustrated below. Each variable will store a particular type of data. In the Fahrenheit-to-Celsius algorithm, the user's input will be a sequence of characters. The **Fahrenheit** and **Celsius** temperatures, on the other hand, will be numbers, possibly with digits after the decimal point. We will need to use the user input names for Fahrenheit, and Celsius, also known as the variables in the Fahrenheit-to-Celsius algorithm. Here's what the algorithm looks like with the variables added:

1. Display a message asking the user to enter a Fahrenheit temperature.
2. Obtain the input entered by the user and store it into user input.
3. Convert user Input into numerical form and store the result into Fahrenheit.
4. Calculate the equivalent Celsius temperature using the formula
   `Celsius = (Fahrenheit-32) x (5/9)`
5. Convert the value of Celsius to a *string format* and display the result.

Programming languages

So far, we now know that algorithms are usually expressed in a natural language like English, with the help of an occasional mathematical formula. That's not good enough for a computer to make most of it though. However, in order to create working programs, we'll need to express our algorithms in a highly precise manner in languages that are specifically designed for computers, popularly known as programming language, compiler or interpreter. The reader will recall that these programming languages were briefly discussed under compiler and programming languages in chapter 7. To fully make sense of programming languages, I have selected visual basic programming language to illustrate our examples. In the sub-topic below we will learn the elements of creating a simple program in Visual Basic.

Creating a Simple Program

In this sub-section we are going to learn how to create a simple addition calculator program in Microsoft Visual Basic programming language. Since this book is not specifically designed for programming, readers are encouraged to consult any visual basic programming book for further reading. It is also important to note that visual basic is an event-driven programming language and as such we are most likely to come across terms like objects, attributes, controls, forms, and others.

Traditionally a typical computer program development will involve four major stages: Analysis, Designing, Coding and Testing.

Analysis

Analysis is the process that involves project planning, gathering the relevant information and defining the problem and system requirements. Here is where programmers become more aware of their actual task. In other words the system specifications would be thoroughly analyzed and made clear to both programmers and the client or the end-user. From my personal experience, the system analysis is the most important process of all software development. Incorrect analysis may lead to technically unsounded and unreliable system and all subsequent efforts would have been wasted.

Designing

The design process is the second stage in software development. Using the product of the analysis, a designer will be able to identify all the major program components including modules, procedures and functions. At this stage a designer would be able to determine whether the proposed project would be feasible or not. The detail steps of programming and the scope of the project would have been uncovered by now. It is at this stage that algorithms, flowcharts and pseudo-codes are derived for each major module. The pseudo-code covers more important details providing a functional and final analysis reflecting the previously defined algorithm and the flowchart. The pseudo-code is more of a literal language expression, which provides an explicit outline to the programming language expressions.

Coding

The coding or programming process is the direct translation of the pseudo-code (a natural language expression) to a specific computer programming language like the C/C++, Visual Basic,

Java, Fortran Pascal, and so on. As in general there are rules governing the expression of any language, the computer programming languages are no exception. In a natural language expression this rule is known as **grammar**. Likewise in computer science each programming language expression is also governed by the special rule known as **syntax**. Wrong syntax will generate an error message. The most important job of a compiler or an interpreter may rest upon the ability to analyze the programming language syntax, before running the program.

Testing

Testing may be the final stage crowning all software programming projects. Usually a testing strategy pre-defined in the process of analysis may be used as the major testing guidelines. Testing is a confirmative process designed to validate the predefined objectives of a programming project. For an example, in the simple addition calculator, the basic testing strategy will likely be as follows:

  a. Test if the user input is calculable.
  b. Test for the size of the user input within the range of a number system.
  c. Test for the size of the calculated results.
  d. Test for the accuracy of the mathematical formula.

Application of Programming Methodology
(Create Simple Addition Calculator Program)

Now using the programming approach discussed above we want to create a simple addition calculator program. In other words, our programming method would embrace all the four stages illustrated above. These stages may include analysis, design, coding and testing.

1. Analysis

Under analysis, we will define the problem and propose a solution at the same time.

A.   Definition of Problem

The current project is focused on the creation of a simple add machine. The program would permit the input of two numbers, and on selecting results or "Add(=)" sign the machine would calculate the sum of the two numbers and display the results in a text box. In the event that the machine was unable to calculate the two numbers an error message would be displayed.

B.   The Proposed Solution to The Defined Problem

Given the above problem, the addition program is a typical example of a **binary** arithmetic operation. This program therefore will require three major variables including **operand#1**, **operand#2** and the **results**. The program will also make the necessary conversions to the appropriate numeric data-type. Since this project is a simple adding machine the value of our result will not be stored permanently. The following global algorithm will reflect the processing of the solution.

Adding Algorithm

```
1.   Begin
2.      Initialize operands
3.      Input Operand#1
4.      Input Operand#2
5.      Calculate Sum(Operand#1, Operand#2 )
6.      Display Results
7.   End
```

2. Design

Using the derived algorithm a pseudo-code can be generated. Here the lines of expressions in a natural language are exploded to accommodate the essential details. Pseudo-code is closer to a programming language. It is important to remember that the pseudo-code can be easily transformed into any programming language.

A. Pseudo-code

a.	**Begin**	
b.	**Do**	
c.		Set Operand#1 -> 0
d.		Set Operand#2 -> 0
e.		Input -> **A**
f.		Input -> **B**
g.		**If A** and **B** = Valid Then
h.		Operand#1= ConvertToNumeric(**A**)
i.		Operand#2= ConvertToNumeric(**B**)
j.		Results = Operand#1 + Operand#2
k.		Display Results
l.	**End if**	
m.	**Until Done**	
n.	**End**	

B. Data Flow Chart

The visual representation of the pseudo-code also known as the **flow chart** can be generated. Some designers generate the data flow chart from the pseudo-code.

Figure 8.9. Data Flow Chart Of A Simple Add Machine

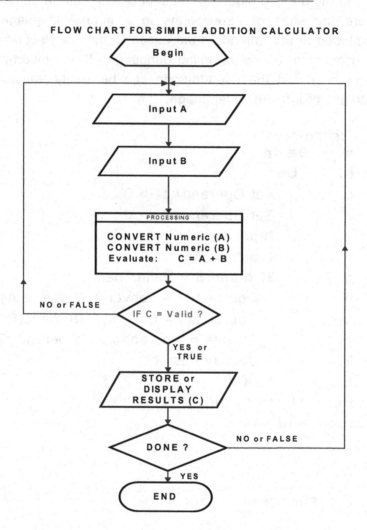

FLOW CHART FOR SIMPLE ADDITION CALCULATOR

3. Coding The Add Machine Program

Once the pseudo-code has been finalized Coding process will begin. The coding process involves the translation of the pseudo-code to a selected computer programming language. At this level the language **syntax** will strictly be respected. In our example, we will choose Microsoft Visual Basic programming language since it's the most popular programming tool. More programming syntax will be discussed further down this chapter, but for now the reader may note that an **apostrophe comma** (′) beginning a statement means a comment. Therefore the compiler will skip that statement during compilation and program execution.

Adding Block Procedure

```
Private Sub cmdEquals_Click()
   Dim Op1 As Single
   Dim Op2 As Single
   'Convert text in Op1 and Op2 to numbers
   Op1 = Val(txtOperand1.Text)
   Op2 = Val(txtOperand2.Text)
   'Two numbers added and reverted to text string
   ' txtResult.Text = Str(Op1 + Op2)
   txtResult.Text=Format(Op1+Op2,"###,###,###.##")
End Sub
```

Since Visual Basic is an event-driven programming language the above block of code by itself is not complete. The interface that allows user interaction with the computer machine is missing. We will therefore introduce a second concept that will further clarify the hidden hitches in the event-driven programming. We will call this part an interface masking. This is where visual objects will be used to present information to guide users, in how to interact with the program.

Programming Concept In Visual Basic

In a simple addition calculator we will need to simulate the ordinary arithmetic method of adding two numbers and displaying the results. Lets assume the adding equation is for example:

Equation (1) . . . .   5 + 4 = 9

The next step is to think computer machine. That is to visualize this equation in terms of a computer machine as shown below. The two most important aspects are **variables** and **values**. The values in equation(1) are **5, 4, 9**. They are called values because they are literals or constants. The variables are labeled **A, B** and **C**. They are variables because their respective values can be changed at anytime during the program execution. In the example

and **C** contains the value **9**.

In the computer machine the variables **A, B,** and **C** are the identification names or labels of memory address. The computer refers to these memory addresses by the variable names.

Variables ⟶	A	B	C
Values ⟶	5	+ 4	= 9

A and **B** are Operands   and **C** is the result.

Putting values into the variables or memory is popularly known as **data input**. And drawing values from the variables or memory is also called **data output**. Now for our calculator machine to work, it will require a declaration of some objects that will represent these variables in the memory.

Visual Basic - Event-Driven Language

In Visual Basic the main object that allows users to interact with the computer machine is called **Form**. A form is an **object** because it is the visual representation of a form **class**. A class is therefore a general definition of a specific type of object, also known as a template. For example **Samuel** is a **boy**. A boy is the name for the class young male persons under the age of eight. Samuel is an **object** since he is a physical representation of the class **boy**. In the object oriented jargon we will also say Samuel is of the class boy (young male person under eight).

Similarly, it will be safer to say that a Form object is an instance of a Form class. A Form like any other object has **attributes** or **properties** and **characteristics** or **methods**. On a form, one can place other smaller objects known in visual basic as **Controls**. Example of controls are Text box or input box (e.g. Text1), Label or display control like the "+ Program" and a Button control called **command button**. Each object has a name and a display title called **caption**. The concept of object-oriented design is very important in today's software development. The beauty of it is the re-usability of programs already created and stored in a special library. The creation of an object from a class is a good example of already created programs called class that can generate an instance called object. To visually represent our simple add calculator will require a form object and sub-objects known as controls already described above.

Table 8.1 A List of Objects and Properties For Add Program

Objects	Properties		Methods
	Name	**Caption**	Event Triggers
Form	**Form1**	**SimpleAddCalc**	
Command	**CmdCalculate**	**Add (=)**	**CmdCalclate**
Text	Text1	Value	
Text	Text2	Value	
Text	Text3	Value	
Label	Label1	Operand#1	
Label	Label2	Operand#2	
Label	Label3	Results	
Label	Label4	+ program	

The program in running mode will look like this:

Figure 8.10. Add Machine Running Mode

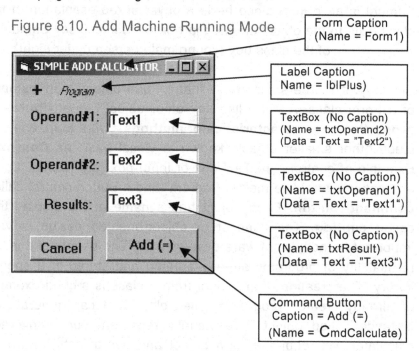

Form Caption
(Name = Form1)

Label Caption
Name = lblPlus)

TextBox (No Caption)
(Name = txtOperand2)
(Data = Text = "Text2")

TextBox (No Caption)
(Name = txtOperand1)
(Data = Text = "Text1")

TextBox (No Caption)
(Name = txtResult)
(Data = Text = "Text3")

Command Button
Caption = Add (=)
(Name = CmdCalculate)

Application of the fundamental concepts discussed above.
After inserting values for the input variables the calculator will
look like the following:

Figure 8.11. The Running Mode of the Add Machine

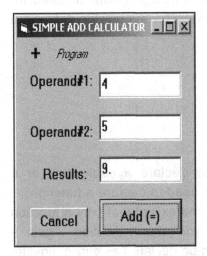 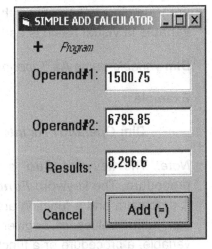

Test no.1   **Test no.2**

How To Store Information

For a program to run, programmers must define variables so they
can be stored in the memory. In Chapter 4, we discussed
memories as temporary storage devices which are very critical to
program execution. In our example we will only calculate numbers
and display the results without storing the values on a permanent
disk.

The Visual Basic programming language is a highly typed
language. Variables must be predefined to determine which types
they belong. A type is a description of data and therefore known
as data type. A data type can be a byte, boolean, character,

string, integer, single, double, currency etc. Such definitions allow the compiler to safely determine the size of memory space needed for each variable. Data type is more important in modern programming languages, so a sub-topic had been set aside to discuss it later in this chapter. The language grammar used for programming is known as the **syntax**. The syntax for defining a variable in visual basic is as follows.

Dim *variable-name* **As** *Typename*

Example:

    Dim *Operand1* As *Integer*

Note: The keyword **Sub** is used to declare a subroutine or a procedure. The keyword ***Function*** is used to declare a function.
A procedure and a function are the same except that only a function can return values of a given data type after execution. Again a variable, a procedure or a function can be declared as public, private, static, or global to promote the access restrictions. Therefore procedures that are triggered from the user-interface interaction must have an *event* function attached to it. An event can be a **Click, KeyPress, KeyUp, KeyDown, Load, GotFocus** etc.

Object Access Concept

A Visual Basic Program is contained in a project (or solution in .NET). The project may carry three major types of modules, Form modules, Standard program modules, and Class modules. All these three modules work together to form one complete application.
Public keyword declaration implies any other object can directly access members of this module from anywhere in the program.

Private keyword declaration implies no other object can directly access members of this module from anywhere in the program except the module itself. The table below shows how Visual Basic project is bundled.

Table 8.2 Presents A List of Object Access Levels In Visual Basic

Module	File Type	Public Access		Private Access	
		IN Module	Outside	IN Module	Outside
Project	*.vbp				
Form	*.frm	Yes	No	Yes	No
Std Module	*.Bas	Yes	Yes	Yes	No
Class Module	*.Cls	Yes	Yes	Yes	No
Sub	-	Yes	No	Yes	No
Function	-	Yes	No	Yes	No

The Block Procedure in the Adding program below.

```
Private Sub cmdEquals_Click()
    Dim Op1 As Single
    Dim Op2 As Single
    'Convert text in Op1 and Op2 to numbers
    Op1 = Val(txtOperand1.Text)
    Op2 = Val(txtOperand2.Text)
    'Two numbers added and reverted to text string
    ' txtResult.Text = Str(Op1 + Op2)
txtResult.Text=Format(Op1+Op2,"###,###,###.####")
End Sub
```

The *Op1* and *Op2* local variables are private to the sub procedure **cmdEquals**. No other sub procedure has access to these variables. On the other hand the form module objects like the *txtResult* and *cmdEquals* are private to the form object but are also public to all member procedures and functions in the form module. This explains for why **txtResult.text** was not explicitly declared in **cmdEquals_Click()** sub procedure but was accessible. The figure 8.12 below illustrates this access restriction concept.

Figure 8.12. Components Constituting a Visual Basic Project

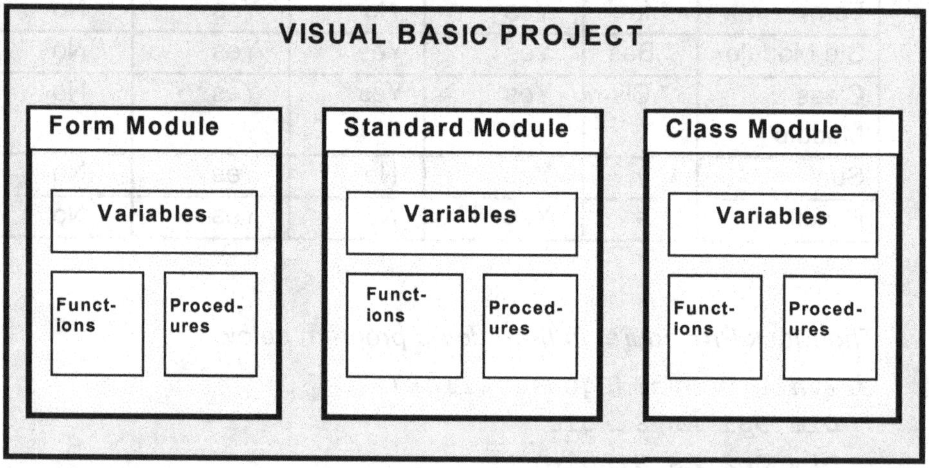

Note: The **Syntax** for a procedure and a function declaration in a standard module would look like the following :

Sub *ProcedureName*([Variable Declarations])
        [Statements]
End Sub

Object access: *ObjectName. Property*

Function *FunctionName*([Variable Declarations]) **As** *TypeName*
   [Statements]
   FunctionName = expression of TypeName
End Function
Function CalculateSquare(**ByVal** anynumber **As** Integer) **As** Integer

```
        Dim intTemp As Integer
        IntTemp = anynumber * anynumber
        CalculateSquare = intTemp
```

End Function

Below is the sample addition calculator program.

```
Private Sub cmdEquals_Click()  'Procedure begins with Sub
    Dim Op1 As Single
    Dim Op2 As Single
    'Convert text in Op1 and Op2 to numbers
    Op1 = Val(txtOperand1.Text)
    Op2 = Val(txtOperand2.Text)
    'Two numbers added and reverted to text string
    'txtResult.Text = Str(Op1 + Op2)
    txtResult.Text=Format(Op1+Op2,"###,###,###.###")
End Sub  'Procedure ends with End Sub
```

Important notes

1. **Objects are called in a program by their names with their properties. Example:** txtOperand1.Text, txtResult.Text, **or** frmCalc.Show. **(txtOperand1= Object; Text= property)**

2. **The Object Command Button with Caption "**Add(=)**" is triggers a** Sub **procedure. Since the procedure is triggered by a mouse click it is automatically named:**

```
cmdEquals_Click().That's ObjectName+_Type of
Event
```

3. The procedure is private and, it is Sub procedure for no value is returned.
4. The values for txtOperand1.Text and txtOperand2.Text are initially string types and are converted to float types before adding them. For example:
Op1=**Val**(txtOperand1.Text) and Op2=**Val**(txtOperand2.Text)
5. The result is reconverted from float type to string type to make it possible to appear in the textbox.
Example: txtResult.Text =**Str**(Op1 + Op2), **or with special format** txtResult.Text=**Format**(Op1+Op2,"###,###,###.##")

General Visual Basic (VB) Language Reference

Under this section, I have included a few visual basic language reference for further reading. These references which I obtained from the Microsoft website will provide an extra learning material to readers who are interested in visual basic programming. I hope by now the reader would be able to create and run visual basic programs from both Salary and Celsius-Fahrenheit algorithm examples discussed earlier in this chapter. We will briefly discuss Data types, Arrays, Loops and a few other key functions in visual basic.

Data type
The characteristics of a variable that determines what kind of data it can hold. Data types include **Byte**, **Boolean**, **Integer**, **Long**, **Currency**, **Decimal**, **Single**, **Double**, **Date**, **String**, **Object**, **Variant** (default), and user-defined types, as well as specific types of objects.

The user-defined types are further defined types using any number of the basic types listed above.

Data Type Summary
The following table shows the supported data types, including storage sizes and ranges.

Table 8.2 The List of Visual Basic Supported Data Types and Sizes

Data type	Storage size	Range
Byte	1 byte	0 to 255
Boolean	2 bytes	**True** or **False**
Integer	2 bytes	-32,768 to 32,767
Long(long integer)	4 bytes	-2,147,483,648 to 2,147,483,647
Single (single-precision floating-point)	4 bytes	-3.402823E38 to -1.401298E-45 for negative values; 1.401298E-45 to 3.402823E38 for positive values
Double (double-precision floating-point)	8 bytes	-1.79769313486231E308 to -4.94065645841247E-324 for negative values; 4.94065645841247E-324 to 1.79769313486232E308 for positive values
Currency (scaled integer)	8 bytes	-922,337,203,685,477.5808 to 922,337,203,685,477.5807
Decimal	14 bytes	+/-79,228,162,514,264,337,593,543,950,335 with no decimal point; +/-7.9228162514264337593543950335 with 28 places to the right of the decimal; smallest non-zero number is +/-0.0000000000000000000000000001
Date	8 bytes	January 1, 100 to December 31, 9999
Object	4 bytes	Any **Object** reference
String (variable-length)	10 bytes + string length	0 to approximately 2 billion
String (fixed-length)	Length of string	1 to approximately 65,400
Variant (with numbers)	16 bytes	Any numeric value up to the range of a **Double**
Variant		Same range as for variable-length **String**

(with characters)	length	
User-defined (using **Type**)	Number required by elements	The range of each element is the same as the range of its data type.

Most of students confuse a class with data structure. As was discussed above *a class is a general definition of a specific type of object, also known as a template.* It is a complete abstract specification of an object. Whereas a data structure can only describe a format of data elements. A class definition may also include data structure known here as attributes or properties; and characteristics or methods. A class definition may constitute a complete program module whereas a data structure may be only partial module. The data structure only describes data types whereas a class definition covers both data types and required programs.

Constant

A **Constant** is a named item that retains a constant value throughout the execution of a program. A constant can be a string or a numeric literal, another constant, or any combination that includes arithmetic or logical operators except logarithms and exponentiation. Each host application can define its own set of constants. Additionally, in VB a constant can be defined by the user with the "**Const** " statement. You can use constants anywhere in your code in place of actual values.

Syntax for declaring a constant:

Const *nameofConstant* **[As** *Typename***] = Value**

Example:

Const *curMinWageRate* **As** Currency **= 5.75**

or

Const *curMinWageRate* **= 5.75**

Also an example of a String constant:

Const *strAuthorsName* = "Bright Siaw Afriyie"

The following are valid word separators for proper casing. They actually represent most important non-printable constants.

Table 8.3 List of Constants for non-printable characters

Description	Constants
Null	Chr$(0)
horizontal tab	Chr$(9)
linefeed	Chr$(10)
vertical tab	Chr$(11)
form feed	Chr$(12)
carriage return	Chr$(13)
space (**SBCS**)	Chr$(32)

The actual value for a space varies by country for **DBCS**

Array
An array is a set of sequentially indexed elements having the same intrinsic data type. Each element of an array has a unique identifying index number. Changes made to one element of an array don't affect the other elements.

Note: Arrays of any data type require **20 bytes** of memory plus **4 bytes** for each array dimension plus the number of bytes occupied by the data itself. The memory occupied by the data can be calculated by multiplying the number of data elements by the size of each element. For example, the data in a single-dimension array consisting of 4 **Integer** data elements of 2 bytes each occupies 8 bytes. The 8 bytes required for the data plus the 24 bytes of overhead brings the total memory requirement for the array to 32 bytes. A **Variant** containing an array requires 12 bytes more than the array alone.

How an array is defined in Visual Basic

Note: An array can be defined in terms of a single or multi-dimensional.

Syntax for single dimension array:

Dim ArrayName(*From Lower Bound* To *Upper Bound*) As Type

Example:

```
        Dim ArrayNumbers(1 To 5) As Integer
Or
        Option Base 1          'important to set first index to 1
        Dim ArrayNumbers(5) As Integer
```

The above statement will create a single dimension array "ArrayNumbers" that can store only 5 numeric elements of an integer type. Below is a visual representation of this array.

Figure 8.13. Visual representation of single-dimension array

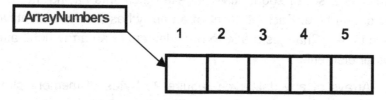

Note: Only values of integer type can be inserted into this array.
To calculate the size of this array we will go through the following four steps:

Step 1: Describe the array as follows

- Dimension of the array   = **Single**
- Number of array elements   = **5**

- Type of each array element = **Integer**

Step 2: Calculate The Overhead

ArrayNumber   = 20 bytes

Dimension Size   = 4 bytes

Step 3: Calculate Size of Elements

No. of Element X Size Of (Type)

5 Elements X Size Of (Integer) =   5 X 2 bytes

= 10 bytes

Step 4: Add Steps 1 through 3   = 34 bytes

Syntax for multi-dimension array:

```
Dim ArrayName(LBound To Ubound,LBound To Ubound)As
Type
```

Example:

```
Dim ArrayNumbers(1 To 5,1 To 3) As Integer
```

Or

Option Base 1   *'important to set first index to 1*
```
Dim ArrayNumbers(5,3) As Integer
```

Below is a visual representation of two dimension array.

Figure 8.14. Visual representation of multi-dimension array

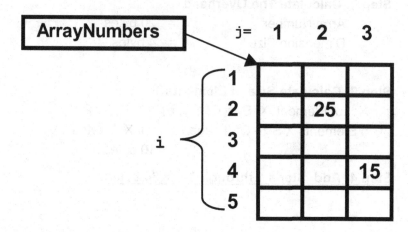

The array has two sets of index variables (i,j). Index *i* ranges from 1 to 5 while index *j* ranges from 1 to 3. Therefore a variable can be stored at any location indicated by **ArrayNumbers(i,j)**. For an example the positions for the values **25** and **15** are indicated as follows:

ArrayNumbers(2,2)= 25 and **ArrayNumbers(4,3)= 15**

Navigating Through Array Elements

This can be achieved through the use of *For* and *Loop* commands. Each iteration of such loops will represent the index corresponding to the position of an array element.

For . . . Next Statement
Do . . . Loop Statement

The **For** and **Do** loops repeat a

is **True** or until a condition becomes **True**.

Syntax:

For . . . Next Statement

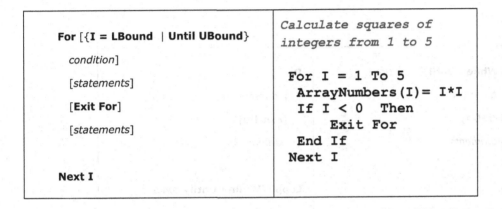

For [{**I = LBound** | **Until UBound**}
 	
 condition]
 	
 [*statements*]
 	
 [**Exit For**]
 	
 [*statements*]
 	
 Next I
	Calculate squares of integers from 1 to 5
	
	 `For I = 1 To 5`
	 `  ArrayNumbers(I)= I*I`
	 `  If I < 0   Then`
	 `        Exit For`
	 `  End If`
	 `Next I`

Figure 8.15. Memory content for a Single Dimension Array

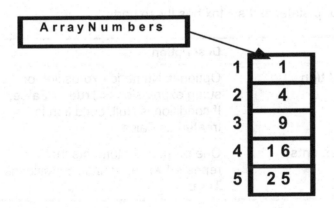

After the execution of the array program, the content of the array *ArrayNumbers* will look like the diagram shown above. The array contains the square values of integers from 1 to 5:
1,4,9,16,and 25

Do . . . Loop Statement

Do [{**While** | **Until**} *condition*]
 	
  [*statements*]
 	
  [**Exit Do**]
 	
  [*statements*]
 	
Loop
	Do
	
	  [*statements*]
	
	  [**Exit Do**]
	
	  [*statements*]
	
	Loop [{**While** | **Until**} *condition*]

The **Do Loop** statement syntax has these parts:

Part	Description
condition	Optional. Numeric expression or string expression is **True** or **False**. If condition is Null, condition is treated as False.
statements	One or more statements that are repeated while, or until, condition is True.

Any number of **Exit Do** statements may be placed anywhere in the **Do Loop** as an alternate way to exit a **Do Loop**. **Exit Do** is often used after evaluating some condition, for example, **If Then**, in which case the **Exit Do** statement transfers control to the statement immediately following the **Loop**.

When used within nested **Do Loop** statements, **Exit Do** transfers control to the loop that is one nested level above the loop where **Exit Do** occurs.

Similarly, the **Do...While/Loop** can be used to iterate the an Array just like the **For...Next loop.**

Do [{**While** | **Until**} *condition*]
 	
[*statements*]
 	
[**Exit Do**]
 	
[*statements*]
 	
Loop
	'Calculate squares of integers from 1 to 5
	
	`I = 0` *'Initialize I to 0*
	Do While I < 5
	`    I = I + 1` *'Increase I by 1*
	`    ArrayNumbers(I) = I*I`
	`    `**If** `I >= 5` **Then**
	`        Exit Do`
	`    `**End If**
	`Loop`

Navigating Through Multi-Dimension Array

The Loop syntax will be nested. This implies there will be loops inserted inside other loops. The depth of nested loop will correspond with the number of dimensions on the array. For example a single dimension array will require a single level of loop. A two-dimension array will require a two-depth level of nested loop. And three-dimension array will require a three-depth level of nested loop.

For i = 1 to n
  For j=1 to m
    [Statement]
  Next j
Next i
	Do
	  j = 0
	  Do
	    [Statement]
	  Loop While j < 3
	Loop While i < 5

This example shows how **Do...Loop** statements can be used. The inner **Do...Loop** statement loops 10 times, sets the value of the flag to **False**, and exits prematurely using the **Exit Do** statement. The outer loop exits immediately upon checking the value of the flag.

```
Dim Check As Boolean
Dim Counter As Integer
Check = True
Counter = 0        'Initialize variables.
Do    ' Outer loop.
   Do While Counter < 20    'Inner loop.
      Counter = Counter + 1  'Increment Counter.
      If Counter = 10 Then     'If condition is True.
         Check = False  'Set value of flag to False.
         Exit Do          'Exit inner loop.
      End If
   Loop
Loop Until Check = False    'Exit outer loop immediately
```

User- Defined Types

The user-defined types are further defined types using any number of the basic types listed above. This is mostly needed when programmers are expected to customize the list of records in a database table. Actually discussion of user-defined types and data structures is beyond the scope of this book. The syntax and examples are shown below to will introduce a basic concept and common usage of user-defined data types

Syntax:

> **Type** *NameofDataStructure*
> [*FieldName* **As** *TypeName*]

> .

> .   [*FieldName* **As** *TypeName*]
> **End Type**

Assuming you wanted to create a data table that contains a list of employees wages as illustrated below:

No	EmployeeID	Name	Wage/US$	Hours
1	0235	Bright A. Siaw	25.75	80
2	5865	Lucy A. Siaw	18.50	80
3	8905	Samuel A. Siaw	12.75	60
4	9085	Joyce K. Siaw	12.85	75

We will now call the data structure **EmployeeRecord** and list of records **EmployeeList.** We will also note that there are five different columns called **data-fields** with assigned names in the very top line of each column. The data field names include **No., EmployeeID, Name, Wage,** and **Hours.** The record number labeled **No.** will be omitted in this process for now. The corresponding user-defined data type will be as shown below:

```
Type EmployeeData
     EmployeeID As String *5
     Name As String * 30
     Wage As Currency
     Hours As Single
End Type
```

Now, we also learned from the data table that there are four records. To accommodate all the four records we will create an *array* of our user-defined type. This array will match the list of records known as the **EmployeeList.**

Dim EmployeeList(1 To 4) As EmployeeData

The array elements for index '1' correspond to the elements in record number '1'. To obtain the elements in record no. 1, you will need to proceed as follows:

```
EmployeeList(1).EmployeeId returns '0235'
EmployeeList(1).Name returns 'Bright A Siaw'
EmployeeList(1).Wage returns 25.75
EmployeeList(1).Hours returns 80
```

Similarly, any of the values in any part of the record list can be changed using the above syntax. For example, the name value **"Bright Siaw Afriyie"**, can be changed to **"Eric Owusuh Siaw"**, as follows:

```
EmployeeList(1).Name = 'Eric Owusuh Siaw'
```

Now what value would the following statement return?

(a) `EmployeeList(2).Name`

(b) `EmployeeList(4).EmployeeId`

(c) `EmployeeList(3).Name`

(d) `EmployeeList(2).Wage`

The remaining records are left for the reader to figure them out.

> *What size of memory space in terms of bytes would be required to store the* **EmployeeList** *array and all its elements?*

Solution

Note:
The sizes of the data-types used in this calculation make reference to Visual Basic To calculate the memory space requirement, the following steps will be used:

a. Calculate the size of **EmployeeData** record a user-defined type
b. The size of the initial array of EmployeeData which is **EmployeeList**
c. The total size of all elements in **EmployeeList** array of records

Step a: The size of a user-defined type is equal to the sum of the sizes of each individual type.

The size of EmployeeData

EmployeeId *(String of length 5)*	=	5 Bytes
Name *(String of length 30)*	=	30 Bytes
Wage *(Currency datatype)*	=	8 Bytes
Hours *(Single datatype)*	=	4 Bytes

Total Size of **EmployeeData**   = **47 Bytes**. . . . (1)

Step b: The size of the array:

```
        EmployeeList(1 To 4) As EmployeeData

Array Overhead              = 20 Bytes
Array Dimension (1)         =  4 Bytes
```
_____

Total Overhead Size   = **24 Bytes** . . . . (2)

Step c: The total size of the 4 array elements.
        (each of **EmployeeData** data-type)

4 X Size of **EmployeeData**   = **188 Bytes** . . . . (3)

The total memory space required to store the array **EmployeeList** is :

```
    ⇨   EmployeeData + Overhead =   (2)+ (3)
    ⇨   Total Values
        =  24 + 188 Bytes

        =  212 Bytes
```

Also as an assignment the reader will have to calculate how much memory space is required to store the above data table.

Below are some useful conversion tools in visual basic programming language.

Table 8.5. Conversion Keyword Summary

Action	Keywords
ANSI value to string.	Chr
String to lowercase or uppercase.	Format, Lcase, UCase
Date to serial number.	DateSerial, DateValue
Decimal number to other bases.	Hex, Oct
Number to string.	Format, Str
One data type to another.	CBool, CByte, CCur, CDate, CDbl, CDec, CInt, CLng, CSng, CStr, CVar, CVErr, Fix, Int
Date to day, month, weekday, or year.	Day, Month, Weekday, Year
Time to hour, minute, or second.	Hour, Minute, Second
String to ASCII value.	Asc
String to number.	Val
Time to serial number.	TimeSerial, TimeValue

Note: Use the **StrConv** function to convert one type of string data to another.

EXERCISE 8.
Answer all questions.

11. List and describe the four major steps involved in software development using the Fahrenheit-to-Celsius conversion method.

12. Why is it important to perform analysis prior to actual programming in a Software project ?

13. What is a data type?. Give two examples and describe their uses.

14. What is an array? Distinguish between a data type and an array. Which one of the two would be best suited for describing a record?

15. Given this expression: `Dim` *arraynames*`(5) As` String `* 20.`
    a. Briefly interpret the expression.
    b. Calculate the size of the *arraynames* with option base 1.
    c. Assuming the initial memory address for the *arraynames* was 7780, calculate the starting address of the third record or index no. 3

16. Refer to the table 8.3 above and answer the following questions:
    a. What is the size of each employee record?
    b. What is the size of array required to accommodate the *Employeelist* carrying 1000 employees?
    c. Calculate the memory address to locate the record bearing the name "Samuel" assuming the beginning address is 5820.

GLOSSARY

AGP	**Accelerated Graphics Port.** This port enables the graphics controller to communicate with the computer and access its main memory. The AGP is a high-speed port designed to handle 3-D technology, and it stores 3-D textures in the main memory rather than the video memory.
Algorithm	Definition of programming outline expressed in a natural language that dictates the order in which the series of instructions would be executed.
ALU	**Arithmetic and Logic Unit.** This is a part of the CPU or microprocessor where the actual computations take place. It consists of integrated circuits which perform arithmetic operations.
API	**Application Programming Interface.** This refers to pieces of programming functions, usually of a lower level design that can be used in creating other programs of a higher level
Array	The area of the RAM that stores the bits. The array consists of rows and columns, with a cell at each intersection that can store a bit.
ASCII	**American Standard Code of Information Interchange.** This is a protocol that formulates the standard character code making it common for all computer machines to interpret. This basic standard coding system can interchange each 7-bits for alphanumeric or non-printable character.
Asynchronous Cache	An **SRAM** that does not require a clock signal to validate its control signals. It is less costly and performance compared to synchronous cache.

ATA	**Advanced Technology Attachment.** ATA is an advanced technology that formulated specifications that defines the IDE disk drive interface in 1980's. *Also see under motherboard devices.*
ATAPI	**Advanced Technology Attachment Packet Interface.** It is a formulated specification that defines the device side characteristics for an IDE connected peripheral like CD-ROM or tape drives and SCSI interface .
Baud	**The data transmission speed measured in terms of symbols per second.** See also *bps*, which stands for bits per second.
Bay	The location inside a computer case where storage systems such as CD-ROM and floppy drives are placed.
BEDO DRAM	**Burst EDO DRAM**: a type of EDO DRAM that can process four memory addresses in one burst. Unlike SDRAM, however, BEDO DRAM can only stay synchronized with the CPU clock for short periods (bursts) and it can't keep up with processors whose buses run faster than 66 MHz.
BIOS	**Basic Input Output System.** The most fundamental software required for any computer to operate is called BIOS, basic input-output system. BIOS is composed of basic instructions needed to startup your PC. These instructions are stored in the ROM (read only memory.
bps	**Bits per second.** The standard data transmission speed measured in terms of bits per second. See also kbps, which stands for kilo-bits per second used for modem speed. Mbps for higher speed transmission lines like the **T1, and T2**. and Gbps.

Burst Mode   Bursting is a rapid data-transfer technique that automatically generates a block of data (a series of consecutive addresses) every time the processor requests a single address. The assumption is that the next data-address the processor will request will be sequential to the previous one. Bursting can be applied both to read and write operations to and from memory.

Bus   Usually, internal data communication channels or cables in the computer.

Cache Controller   **Cache Controller.** The circuit in control of the interface between the CPU, cache and DRAM - main memory

Cache Hit   When the address requested by the CPU is found in cache. Conversely, cache miss is when its not found.

Cache Memory   **Cache Memory.** A small block of high-speed memory (usually SRAM) located between the CPU and main memory that is used to store frequently requested data and instructions. Properly designed, a cache improves system performance by reducing the need to access the system's slower main memory for every transaction.

CD-ROM   **Compact Disc, read-only memory.**
A storage device that reads CD-ROMs. These drives are standard on many PCs and are available in various speeds, which are represented as multiples of X. Most systems presently offer drives averaging 24X or 52X.

CD-RW   **compact disc, rewrite-able drive.**
A CD-based drive that can record or erase data, as well as read it.

CGA   **Color Graphic Adapter.**
It is basic colors display component that succeeded monochrome adapters.

Check Bits

Check Bits. Extra data bits provided by a DRAM module to support ECC function. For a 4-byte bus, 7 or 8 check bits are needed to implement ECC, resulting in a total bus width of 39 or 40 bits. On an 8-byte bus, 8 additional bits are required, resulting in a bus width of 72 bits. Sometimes called parity bit.

Chip

An integrated circuit. A thin piece of silicon that contains all the components of an electronic circuit

Chipset

A collection of microchips that allow hardware peripherals in PCI and ISA buses to communicate without using the CPU. This frees processing time in the CPU to perform other tasks.

CISC

CISC *stands for* Complex Instruction Set Computer. It is originally Intel brand computer architecture in which the CPU uses micro-code to execute a very comprehensive instruction set. *Also see* RISC.

Cluster

A block of space on PC storage media where the operating system identifies as the smallest logical units used for data storage. In DOS platform a cluster consists of 15 sectors on a hard disk.

CMOS

CMOS *stands for* Complementary Metal Oxide Semiconductor. It controls the memory that stores configuration jumper settings of a computer. This is a process that uses both N- and P-channel devices in a complimentary fashion to achieve small geometries and low power consumption.

Connectivity

The establishment of communication links between two or more computers that are directly or indirectly connected in a network for the purpose of data exchange.. It usually requires a modem or network interface card (NIC) in each end to establish a connection. Example is Internet access

CPU	**Central Processing Unit.** Think of the CPU, or microprocessor, as the brain of every system. The CPU is a silicon chip that deciphers and initiates your commands. The clock speed of CPUs(recorded in Megahertz - millions of cycles per second, or Gigahertz), is a major factor in how fast the microprocessor can perform its calculations. Nowadays some CPU's run at 3.2GHz plus
CRT	**CRT** *stands for* Cathode Ray Tube. CRT is the technology used to manufacture computer monitors other than the flat LCD and Gas plasma. Display is achieved by passing a beam of cathode ray electrons to light up the pixels/dots on screen and colors achieved by combining the standard colors Red-Green-Blue.
Cylinder	**A Cylinder** is the sum of the total areas of a track and corresponding tracks of the same position on all the platters of a storage disk. For example, in a hard disk that has 4 platters with 10 tracks each. There are therefore 10 cylinders composed of 8 tracks each.
Data	Data comprises any set of information that can be manipulated in a computer system.
DDR	**DDR.** Double Data Rate: a memory technology that works by allowing the activation of output operations on the chip to occur on both the rising and falling edge of a clock cycle, thereby providing an effective doubling of the clock frequency without increasing the actual frequency.
Desktop	Desktop refers to a personal computer which system unit is usually placed with the large side on top of a computer desk and at the same time serving as a support stand for its monitor.
DIMM	**Dual in-line memory module.** A circuit board that contains multiple memory chips that can be inserted into memory expansion slots to increase RAM. DIMM supports 64-bit and higher buses and have 168 pins.

Disk Drive	A mechanism that holds, spins, reads, and writes either magnetic or optical disks
DMA	**DMA** *stands for* **Direct Memory Access.** This is a method of an interface memory access without passing through the processor.
DOS	**DOS** *stands for* **Disk Operating System.** The original operating system for personal computer. Dos has now evolved into windows or GUI operating system.
dpi	**dpi** *stands for* **Dot Pitch per Inch.** The smaller dots in monitors usually determine the sharpness of graphics. The smaller the dot pitch the higher the dot pitch per inch (dpi) value, the higher the sharpness of graphics displayed
DRAM	**Dynamic random-access memory.** A type of computer memory that is popular in PCs because of its low price.
DRDRAM	**Direct Rambus DRAM**. It is a totally new and a very fast DRAM that operates by actually narrowing the bus path and treating the memory bus as a separate communication channel. Its RAM architecture is complete with bus mastering (the Rambus Channel Master) and a new pathway (the Rambus Channel) between memory devices (the Rambus Channel Slaves). A single Rambus Channel has the potential to reach 500 MBps in burst mode; a 20- fold increase over DRAM.
DVD	**Digital Video Disc drive.** A drive that can read audio and video software CD-ROMs and DVDs, which store up to 4.7 gigabytes (GB) of information on each side of the disc.
EBCDIC	**Extended Binary Coded Decimal Interchange Code.** It is an extension of the ASCII code. EBCDIC is an 8-bit character code that is used with most computer equipment manufactured by the IBM company.

EDO DRAM	**Extended Data Out Random Access Memory**. A form of DRAM that has a two-stage pipeline, which lets the memory controller read data off the chip while it is being reset for the next operation. While similar in performance to synchronous DRAM (SDRAM), it cannot support bus speeds above 66MHz.
EDRAM	**Enhanced Dynamic Random Access Memory**. A form of DRAM that boosts performance by placing a small complement of static RAM (SRAM) in each DRAM chip and using the SRAM as a cache. Also known as cached DRAM, or CDRAM.
EEPROM	**Electrically Erasable Programmable Read Only Memory**. A special type of read-only memory (ROM) that can be erased and written electrically. EEPROM maintains its contents without power backup and is frequently used for system-board BIOS's.
Embedded processor	A chip designed with a specific set of usable instructions.
EPROM	**Erasable Programmable Read Only Memory**. An integrated circuit memory chip that can store programs and data in a non-volatile state. These devices can be erased by high-intensity ultraviolet (UV) light and then rewritten, or "reprogrammed", in a manner similar to common DRAM. EPROM chips normally contain UV-permeable quartz windows exposing the chips' internals.
Ethernet	The most common type of protocol used for local-area networks. Protocols are sets of standards that spell out the rules for how PCs communicate and exchange data.
Expansion cards	Circuit boards that fit into expansion slots on the motherboard to provide the computer with new devices, such as a graphics card, modem, and scanner.
Expansion slot	A connector or an available opening on the motherboard designed to allow the addition of electronic cards or circuit boards. The standard expansion slots include PCI, AGP, or ISA connections. Expansion cards fit in expansion slots.

FAT	**File Allocation Table.** An operating system concept developed by Microsoft to manage files in IBM compatible personal computers. Common examples FAT16 and FAT32, 16 and 32-bit FAT systems. **NTFS** is another type of file system concept.
FDD	**Floppy disk drive.** Storage read-write device which uses a floppy disk as the storage media. Unlike the hard disk drives the floppy drive is separable from its removable read-write floppy disk also known as diskette.
Flash Memory	Flash memory is a non-volatile memory device that retains its data when the power is removed. The device is similar to EPROM with the exception that it can be electrically erased, whereas an EPROM must be exposed to ultra-violet light to erase.
FPM DRAM	**Fast Page Mode RAM.** This is a timing option that permits several bits of data in a single row on a DRAM to be accessed at an accelerated rate. Fast Page Mode involves selecting multiple column addresses in rapid succession once the row address has been selected.
GB	**Gigabyte.** A Gigabyte, also referred to as a "**Gig**", is a measurement of computer storage space that equals 1,073,741,824 bytes or 1 billion characters. Bytes are typically represented in computer terminology by an uppercase "B" **TB =Terabyte. And it is equal to 1024 x 1 Gigabyte.**
GHz	**GHz** *stands for* **Gigahertz.** A measurement used to gauge the speed of a CPU. One gigahertz is equivalent to 1 billion cycles per second.
Graphics	A video adapter that has its own processor to enhance performance. These accelerators are designed to handle high-end graphic activities, leaving the CPU to handle other system requirements. Accelerators typically use conventional DRAM, or a special type of video RAM (VRAM), to accommodate the video circuitry and the processor to access the memory simultaneously.

GUI	**GUI** *stands for* **Graphical User Interface.** Also pronounced as "**Gooey**". An interface that displays information like windows with graphical representation of objects.
Hardware	The physical parts of the computer system, keyboard, monitor, and computer case.
HDA	**Hard Disk Assembly.** The packaging of modern hard drives in which the disk and drive are fused together as one main device that can store and retrieve information. These drives are sealed boxes typically found inside the PC case.
HDD	**Hard Disk Drive.** The main device a computer uses to permanently store and retrieve information. These drives are sealed boxes typically found inside the PC case. Today's hard drives have an average storage capacity of 10 gigabytes to over 300 gigabytes.
HPM	Hyper Page Mode: in DRAM operation, another term for EDO or Extended Data Out.
IC	**Integrated Circuit.** A complete circuit or a chip, built by a chip fabrication process. The actual circuitry component that constitutes the chip.
Index	**Index.** In memory, an index is the subset of the CPU address bits used to get a specific location within cache.
Information Processing	Capturing, storing, updating, and retrieving data and information.
Input	The data that is entered into a computer.
Interleave	The process of taking data bits (singly or in bursts) alternately from two or more memory pages (on an SDRAM) or devices (on a memory card or subsystem).
I-P-O	**Input-Process-Output** method to process data.

IRQ	**Interrupt ReQuest.** It a setup that allow the operating system to interrupt a Process in progress in order to accommodate another process of higher priority. For example a key press.
ISA	**Industry Standard Architecture.** A standard bus model that supports 16-bit expansion cards. ISA buses are slowly being displaced in systems by faster PCI and USB buses because the majority of hardware peripherals are now constructed for these newer standards.
KB	**KB** *stands for* **Kilobyte.** An amount of storage equivalent to 1,024 bytes, or about 1,000 characters of information.
Keyboard	A computer input device that uses a set of keys to put data into the computer. It is the main input device for a computer that is similar to an electric typewriter keyboard with a few extra keys for various functions.
L1 Cache	**Level 1 Cache.** Also known as *primary cache* is a type of faster memory that stores information that is frequently accessed
L2 Cache	**L2 Cache.** A type of memory that is external to the microprocessor but provides quicker access than the primary cache. You will typically find L2 cache on a separate chip, except for the Pentium Pro, which has an L2 cache on the same chip as the microprocessor.
LCD	**Liquid Crystal Display.** This is a kind of monitor usually a flat screen using crystal display technology. Examples are Flat monitors, laptops and Palm Pilot and CE devices.
LSI	**Large Scale Integration.** A classification of the CPU as measured by the number of the integrated circuits (**ICs**) that are linked together, usually 3,000 IC's or less. For example CPU's below PII are mostly **LSI** categories. PII upwards are **VLSI**. Also *see* **VLSI**.

LTO	**Linear Tape Open.** This is a kind of a backup tape technology used for data safeguard. The following are other exa Advanced Intelligent Tape.
Mb	**megabit (Mb).** A measurement equivalent to approximately 1 million bits. A bit is a binary unit. Bits are typically represented in computer terminology by a lowercase "**b**."
MB	**Megabyte.** A megabyte, also referred to as a meg, is a measurement of computer storage that equals 1,048,576 bytes. Bytes are typically represented in computer terminology by an uppercase "**B**"
MB/Sec	**MB/Sec.** The interface speed of a hard disk drive. It is also the transmission speed for the hard disk drive to move data across the ribbon cables in the computer. *See* rpm
MDA	**Monochrome Display Adapter.** This is an adapter installed in early computers that can only display one color.
Memory	A place to store information-RAM, ROM or Disk. Also **see** RAM.
Memory Bank	A logical unit of memory in a computer, the size of which the CPU determines. For example, a 32-bit CPU requires memory banks that provide 32 bits of information at a time. A bank can consist of one or more memory modules.
Memory Controller	An essential component in any computer. Its function is to oversee the movement of data into and out of main memory. It also determines what type of data integrity checking, if any, is supported.
Memory Cycle	Minimum amount of time required for a memory to complete a cycle such as read, write, read/write, or read/modify/write.

MHz	**Megahertz.** A measurement used to gauge the speed of a CPU. One megahertz is equivalent to 1 million cycles per second.
Microprocessor	An integrated circuit known as the central processing unit, or CPU, that controls the entire computer system. It coordinates all of the actions of the computer carrying out instructions, performing calculations, and interacting with all the components.
MNOS	Metal Nitride Oxide Semiconductor: the technology used for EAROMs (Electrically Alterable ROMs); not to be confused with NMOS.
Modem	**Modem.** Acronym for *mo*dulator/*de*modulator. A communications device that allows a computer to transmit data over analog telephone lines. The present standard speed of modems is 56Kbps.
Monitor	A computer output device that uses a display screen to present the processed information. A type of display screen for desktop systems that resembles a TV screen in a hard shell case. Monitors attach via cables to a computer and display the image signals produced by the system.
MOS	Metal-Oxide-Semiconductor: layers used to create a semiconductor circuit. A thin insulating layer of oxide is deposited on the surface of the wafer. Then a highly conductive layer of tungsten silicide is placed over the top of the oxide dielectric.
Motherboard	The main circuit board, housing the microprocessor and providing the means of connecting all the components that make up the computer. It is the main circuit board inside a computer that provides the foundation for the system. The motherboard holds all the internal circuitry for the system, such as the CPU, buses, memory sockets, and expansion slots.

Mouse Computer input device used as a pointing and drawing instrument by selecting specific positions on the monitor display. A device that provides the user with an on-screen control in a graphical user interface.

NIC **Network Interface Card**—An expansion board that allows the computer to be hooked to a network.

NMOS N-channel Metal Oxide Semiconductor: pertains to MOS devices constructed on a P-type substrate in which electrons flow between N-type source and drain contacts. NMOS devices are typically two to three times faster than PMOS devices.

Operating System Software that handles the computer's basic functions and acts a foundation to run additional programs. Operating systems recognize keyboard input, send the output to monitors and printers, control peripheral devices, handle system security, and keep records of files and directories. Microsoft's Windows 95, Windows 98, and Windows NT are examples of operating systems.

Output The computer-generated information that is displayed to the user in discernible form on screen, printer, or speaker.

Page On a DRAM, the number of bits that can be accessed from one row address. The size of a page is determined by the number of column addresses. For example, a device with 10 column address pins has a page depth of 1024 bits.

PCI **Peripheral Component Interconnect**—An Intel-designed bus that features quick communications between a peripheral and the computer's CPU. PCI buses allow additional peripherals to be installed on a system, and they provide plug-and-play capability.

Personal Computer A microcomputer that serves one user at a time. It's commonly known as a PC.

Pipeline
In DRAMs and SRAMs, a method for increasing the performance using multistage circuitry to stack or save data while new data is being accessed. The depth of a pipeline varies from product to product. For example, in an EDO DRAM, one bit of data appears on the output while the next bit is being accessed. In some SRAMs, pipelines may contain bits of data or more.

Pipeline Burst Cache
A type of synchronous cache that uses two techniques to minimize processor wait states - a burst mode that pre-fetches memory contents before they are requested, and pipelining so that one memory value can be accessed in the cache at the same time that another memory value is accessed in DRAM.

Pixel
Picture Element. A unit of measuring picture resolution. For example a resolution of **800 x 600** will equal **480,000 pixels**

Ports
Multi-pin connectors that enable external devices, such as monitors and printers, to be hooked to the computer.

Pseudo-code
A mixture of natural language and a programming language. A more detailed expression of algorithm closer to a programming language outlined in a natural language that dictates the order in which the series of instructions would be executed. Programmers simply translate the expression of pseudo-code into a programming language. This process is known as coding.

RAM
Random-Access-Memory— Temporary storage memory chips that form the computer's primary workspace. Its content is lost if power is disrupted; the computer uses RAM to hold information it is processing. The amount of RAM determines the size and number of programs that can be used simultaneously and affects the overall processing speed of the computer.

A term that refers to the amount of sharpness and clarity of an image. Higher resolution devices produce sharper, and more highly defined images.

RIMM
A form of chip packaging that is similar to DIMMs to be used with the next generation of Direct DRAM memory subsystems.

RISC
RISC *stands for* **Reduced Instruction Set Computer**. A kind of a CPU architecture which keeps its instruction sets constant, banning indirect addressing mode and retaining only the instructions that can be overlapped and made to execute in one machine cycle or less. Also *see* **CISC**.

ROM
Read Only Memory - Memory chips that have their stored content entered at the time of fabrication. It can be written to only once, and content is **not** lost if power is disrupted.

RPM
Rotation Per Minute – A measuring unit for the rotational speed of a hard disk drive. Also see the interface speed of a hard disk drive - **MB/Sec.**

Scanner
Computer input device that can read text, images, and bar codes and translates them into digital code.

SDRAM
Synchronous DRAM—A memory type that provides higher speeds than DRAM by synchronizing the computer's internal clock with the memory, which allows for speeds up to 100MHz.

Secure Digital
SD. It is a postage stamp size portable flash memory format developed by Toshiba, Sandisk and Panasonic. Content encoded on an **SD** card may be encrypted, providing copyright protection of intellectual properties. Expected to the industry standard for the warehousing and transfer of digital media including music, still and moving video, talking books, etc.

SGRAM

Synchronous Graphics RAM —A type of DRAM memory used in conjunction with graphics accelerators and video adapters.

SIMM

Single In-Line Memory Module. On Pentium-class PCs, SIMM-style RAM chips replaced the dual in-line package (DIP) chips, identifiable by two rows of protruding legs, that were popular in the 1980s. They are themselves being replaced by the DIMM module.

Software

A series of instructions usually known as programs that provide direction to perform a particular task. Unlike the hardware the software is intangible, but an intellectual property.

Sound card

An expansion board that provides audio capability and enables the computer to digitize sound beyond basic beeps.

Universal Serial Bus (USB) They are also known as ports that make adding hardware peripheral, such as a printer, scanner, or joystick, a snap. An external bus expected to eventually replace serial and parallel ports for adding peripherals to a system. Most of today's PCs feature two or more USB ports that provide plug-and-play and hot-swapping capabilities.

VGA

Video Graphics Adapter. This device permits a computer to display up to 4-bits or 16 colors.
SVGA - Super VGA displays 8-bits or 256 colors
XGA   - Extra-Super VGA displays 16-bits to 24-bits of colors.

VLSI

Very Large Scale Integration. A classification of the CPU as measured by the number of the integrated circuits (**ICs**) that are linked together, usually between 3,000 and 100,000 IC's or less. For example CPU's like PII and above PIII, PIV are mostly **VLSI** categories. PII downwards are **LSI.** Also *see* **LSI.**

VRAM   **Video Random Access Memory**: a dual-ported DRAM designed for graphics and video applications. One port provides data to the CRT, while the other is used for read/write transfers from the graphics controller. See also **WRAM.**

WRAM   Windows Random Access Memory: a form of VRAM used exclusively by Matrox Graphics. WRAM has added logic designed to accelerate common video functions such as bit-block transfers and pattern fills. It can substantially speed up certain graphical operations such as video playback and screen animation.

A type of storage system designed by Iomega Corp. that holds up to 100MB of data on portable diskettes. Zip drives are a popular device used for storing, transporting, and backing up files.

ANSWERS

Chapter One : Exercise 1

1. E	11. D
2. A	12. A
3. B	13. A
4. D	14. B
5. A	15. D
6. E	16. D
7. B	17. B
8. C	18. D
9. B	
10. D	

Chapter Two : Exercise 2

1. B	11. C
2. D	12. A
3. B	13. C
4. E	
5. A	
6. A	
7. B	
8. A	
9. A	
10. E	

Chapter Three : Exercise 3

1. B	11. A
2. A	12. E
3. C	13. A
4. B	14. A
5. A	15. C
6. B	16. A
7. B	17. A
8. A	18. E
9. B	19. E
10. A	20. B

Chapter Four : Exercise 4

1. E	11. A	21. B
2. D	12. B	22. C
3. A	13. B	
4. C	14. C	
5. B	15. A	
6. B	16. D	
7. E	17. C	
8. A	18. E	
9. E	19. B	
10. D	20. E	

ANSWERS

Chapter Five : Exercise 5			Chapter Six : Exercise 6	
1.	11. A	21. E	5. C	15. B
2.	12. E	22. B	6. E	16. D
3.	13. D	23. E	7. D	17. D
4.	14. A	24. B	8. E	18. A
5.	15. B	25. C	9. E	19. B
6.	16. B		10. A	20. D
7.	17. C		11. C	21. B
8.	18. A		12. A	22. A
9.	19. D		13. B	23. D
10. E	20. C		14. D	24. B
				25. B

Index

A

B

Index

Printed in the United States
By Bookmasters